W9-BUI-581

Native Americans in Sports

Volume 1

Native Americans in Sports

Volume 1

Edited by C. Richard King

SHARPE REFERENCE
an imprint of M.E. Sharpe, Inc.

SHARPE REFERENCE

Sharpe Reference is an imprint of M.E. Sharpe INC.

M.E. Sharpe INC.
80 Business Park Drive
Armonk, NY 10504

Library of Congress Cataloging-in-Publication Data

Native Americans in sports / C. Richard King, editor.
p. cm.
Includes bibliographical references and index.
ISBN 0-7656-8054-8 (set : alk. paper)
1. Indians of North America—Sports—Dictionaries. 2. Indian athletes—Biography—Dictionaries. 3. Indians of North America—Societies, etc.—Dictionaries. 4. Sports—United States—Societies, etc.—Dictionaries. I. King, C. Richard, 1968– II. Sharpe Reference (Firm)

GV583.N34 2003
796'.089'97—dc21

2002042800

Cover photo credits: Jim Thorpe, Notah Begay, and Johnny Bench provided by AP/Wide World Photos; Bryan Trottier (© Bruce Bennett/Bruce Bennett Studios); female runner (© Alan Bailey/Rubberball Productions/PictureQuest); lacrosse player (© Corbis)

Printed and bound in the United States of America

The paper used in this publication meets the minimum requirements of American National Standard for Information Sciences—Permanence of Paper for Printed Library Materials, ANSI Z 39.48.1984.

MV (c) 10 9 8 7 6 5 4 3 2 1

CONTENTS

VOLUME 1

VOLUME 2

TOPIC FINDER

INDIVIDUALS

SPORTS

TEAMS

INSTITUTIONS AND ORGANIZATIONS

KEY PERSONNEL, CULTURAL THEMES, AND SOCIAL ISSUES

EDITOR

C. Richard King
Washington State University

CONTRIBUTORS

James Allen

Lawrence R. Baca
National Native American Bar Association

William J. Bauer, Jr.
University of Wyoming

Yale D. Belanger
Trent University

Gregory Bond
University of Wisconsin

Larry S. Bonura
Independent Scholar

John P. Bowes
University of California at Los Angeles

Wilburn C. Campbell
Miami University

Brian S. Collier
Arizona State University

Cary C. Collins
Independent Scholar

Kathy Collins
Washington State University

Rory Costello

Scott A.G.M. Crawford
Eastern Illinois University

Wade Davies
University of Montana

Rick Dyson
Lamar University

Victoria Edwards
Dalhousie University

Eric Enders
Triple E Productions

Lisa A. Ennis
Austin Peay State University

Donald M. Fisher
Niagara County Community College

Janice Forsyth
University of Western Ontario

M. Todd Fuller
Pawnee Nation of Oklahoma

Audrey R. Giles
University of Alberta

Doron Goldman
University of Massachusetts at Amherst

Anne Wheelock Gonzales
Wings of America

Barbara A. Gray
Arizona State University

Adolph Grundman
Metropolitan State College of Denver

Dixie Ray Haggard
University of Kansas

M. Ann Hall
University of Alberta

Roger D. Hardaway
Northwestern Oklahoma State University

Othello Harris
Miami of Ohio University

Susan Haslip
Carleton University

Edward W. Hathaway
Rational Pastimes Sports Research Services

Beth Pamela Jacobson
University of Connecticut

Janis A. Johnson
University of Idaho

Vickey Kalambakal
Arizona State University

Susan Keith
Angelo State University

Penelope M. Kelsey
Rochester Institute of Technology

CONTRIBUTORS

Kay Koppedrayer
Wilfrid Laurier University

Dale E. Landon
Indiana University of Pennsylvania

Pat Lauderdale
Arizona State University

Rita M. Liberti
California State University at Hayward

Bill Mallon
Independent Scholar

William R. Meltzer
Independent Scholar

Diana Meneses
Arizona State University

Christine O'Bonsawin
Western Ontario University

Caoimhín P. Ó Fearghail
Northern Arizona University

Royse Parr
Independent Scholar

David Porter
William Penn University

Jeffrey Powers-Beck
East Tennessee State University

Robert Pruter
Lewis University

SuAnn M. Reddick
Independent Scholar

Edward J. Rielly
Saint Joseph's College of Maine

Mike Robbins
Freelance Journalist

Nicolas Rosenthal
University of California at Los Angeles

Frank A. Salamone
Iona College

Kelly Boyer Sargent
Freelance Writer

Raymond Schmidt
Independent Scholar

Michael Sherfy
University of Illinois at Urbana-Champaign

Ronald A. Smith
Pennsylvania State University

T. Jason Soderstrum
Iowa State University

Charles Fruehling Springwood
Illinois Wesleyan University

Ellen J. Staurowsky
Ithaca College

Glenn Ellen Starr Stilling
Appalachian State University

David Hurst Thomas
American Museum of Natural History

Richard Thompson
Independent Scholar

Grace F. Thorpe
Native American Activist and Author

Rebecca Tolley-Stokes
East Tennessee State University

John Valentine
MacEwan College

Michel Vigneault
Université de Québec

J.R. Wampler
Miami University

Kevin B. Wamsley
University of Western Ontario

Lisa R. Williams
Washington State University

Kevin Witherspoon
Florida State University

INTRODUCTION

The first Americans were surely the continent's first athletes as well. In each of the several hundred societies native to North America, men and women engaged in sporting activity. Lacrosse may be the indigenous sport most familiar to contemporary Americans and Canadians, but Native nations took part in numerous other contests, including running, ball games, horse races, archery, and wrestling. These events often served spiritual, social, political, and recreational functions, weaving ceremonial, community, and competitive features into the fabric of sport. Beginning in the sixteenth century, social interactions, cultural conflicts, and governmental policies profoundly altered the patterns of play. In the wake of Euro-American colonization, marked by assimilation, dispossession, and the eradication of indigenous institutions, including sport, Native Americans have embraced athletics. Under often oppressive circumstances, they have enjoyed, reinterpreted, and excelled at modern sports. They have at the same time endeavored to reclaim more traditional forms of athletic expression, making powerful statements about survival, sovereignty, identity, and community.

Forgotten Heroes

Although countless great indigenous athletes have played sport in North America, most people in the United States and Canada know little about them. Racial stereotypes, myopic media coverage, the isolation of reservation communities, and the dominance of a handful of revenue-producing sports have combined to make Native Americans all but invisible in the world of sports. In fact, it is likely that the vast majority of fans can name only one or two superstars—most likely James Francis Thorpe, who distinguished himself at Carlisle Indian School as an All-American in football and at the 1912 Olympics, where he won the pentathlon and decathlon. He went on to careers in professional football and baseball, then served as founding president of the National Football League (NFL) and pursued a brief film career. After his death he was the unanimous selection as the greatest athlete of the first half of the twentieth century and was among those deemed to have been the greatest in the entire century.

The achievements of Thorpe make it all the more difficult for the public to recognize and appreciate the myriad other Native Americans who have excelled at sports. To name only a few of the representative indigenous superstars of modern sport: Louis Tewanima and William Mervin "Billy" Mills in running; Charles Albert "Chief" Bender, Allie Pierce "Super Chief" Reynolds, and John Meyers in baseball; Joseph Napleon "Joe" Guyon, James William Plunkett, Jr., and William "Lone Star" Dietz in football; Ron Delorme, Bryan "Trots" Trottier, and George E. Armstrong in hockey; Angelita Rosal Bengtsson in table tennis; Danny "Little Red" Lopez in boxing; Corey Witherill in auto racing; Ross Anderson in speed skiing; and Rod Curl, and more recently Notah Ryan Begay III, in golf. In fact, the public is often unaware of the Native heritage of many athletes, such as baseball players Johnny Lee Bench and Wilver Dornel "Willie" "Pops"

Stargell, runner Patty Catalano, college basketball coach Kelvin Sampson, basketball player Cherokee Bryan Parks, golfer Eldrick "Tiger" Woods, and the owner of the Tennessee Titans, Kenneth Stanley "Bud" Adams, Jr.

In addition to these great individuals, a number of all-Indian teams once rose to prominence. In the early years of the twentieth century, the exploits of the teams at boarding schools, particularly Carlisle Indian Industrial School and Haskell Institute, dazzled Americans. At the same time, barnstorming baseball teams, like Green's Indians, excelled as both athletes and entertainers. The Oorang Indians, a short-lived professional football franchise captained by Jim Thorpe, strove to do the same, staging stereotypical Indian dances, dressing in feathers, and putting on displays of skill during the half-time of their games. Later, the Hominy (Oklahoma) Indians football team briefly captured the public imagination, besting nearly all opponents, including the Cleveland Browns and the New York Giants of the National Football League.

Failure to remember the contributions of indigenous peoples obscures a rich heritage and offers a distorted portrait of athletics in North America. In recalling great, if too often forgotten, players and games, *Native Americans in Sports* strives to offer a more complete picture of American sporting worlds.

Heritage

Native Americans have always played sports. The exact games played varied but included lacrosse and archery, shinny and snowsnake, running and archery, snowshoe and canoe races, and double ball and toli. In general, across North America sports were not played for mere amusement. Rather, they were deeply rooted in indigenous worldviews and everyday life. Joseph Bruce Oxendine has identified several key features uniting traditional forms of athletic expression in Native America: deep ties between sports and other aspects of life, particularly the social and spiritual; intense attention to physical training and spiritual rites for individuals and their communities in advance of sporting events; a general lack of standardization and quantification; separate competition for men and women; emphasis on fair play; and the importance of gambling. In addition, indigenous sporting worlds stressed the spiritual and social utility of sport while downplaying rules and winning.

Little is known of the earliest athletes. In fact, only in the nineteenth century do glimpses of individual Native American athletes begin to appear, first in accounts of indigenous societies and then, more importantly, in association with the rise of modern sport. George Caitlin was so taken with the play of Tullock-chish-ko (He Who Drinks the Juice of the Stone) that he described him as "the most distinguished ball-player in the Choctaw nation," painting a full portrait in his game attire. Later, others would comment on the greatest of indigenous runners like Chief Big Hawk (who reportedly ran a sub-four-minute mile).

Assimilation

Native Americans played a starring role in sports as they became increasingly institutionalized, modernized competitions for public consumption. In running, for instance, American Indians captivated thousands. A decade after Senneca John Steeprock had attained notoriety, Louis "Deerfoot" Bennett began competing in footraces in the Northeast, and in the early 1860s he became a sensation on both sides of the Atlantic. It was not until after the bulk of Native Americans were settled on reservations, however, that they became more active, even integrated, in Euro-American sporting worlds. In fact, the federal boarding school system established in the last quarter of the nineteenth century became an important training ground for many Native American athletes. Sports became increasingly important elements of such schools as means of assimilating American Indians.

Sport fit nicely within the Euro-American program to assimilate indigenous peoples through education, to "kill the Indian and save the man." Consequently, indigenous games, contests, and traditional forms of play, like so much else, were suppressed and replaced by forms deemed proper and civilized by Euro-American reformers. Physical education was a key component of boarding schools and later was supplemented by organized, interscholastic athletic competitions. In these varied forms, educators hoped to instill a competitive spirit, discipline, morality, and manliness. In time, they would come to see sports as a powerful public relations tool as well.

Importantly, athletics proved to be something other than a one-way street toward civilization. It offered Native Americans important occasions to express and define themselves, to have fun, and to resist the very regimes Euro-Americans sought to impose upon them. Others, like Charles Eastman discussed recognized sport as an important means of racial uplift. For all of the rhetoric, many indigenous athletes had more practical concerns, more material motivations; sports proved a great way for some to make livings and names for themselves.

Declining Fortunes

While indigenous athletes were prevalent in North American sports up through the first third of the twentieth century, a number of social shifts would alter their prominence. At least seven fundamental changes explain the decline of the indigenous athlete: the closing of Carlisle and Haskell (still open as a tribal college); the shortage of Indian institutions of higher education; substandard local schools; growing resistance to assimilation; efforts, known as "termination," to abrogate rights, responsibilities, and relationships between the federal government and recognized tribes; the civil rights movement and the associated reinterpretation of black-white race relations, which in turn eclipsed the "Indian problem" and its significance for most Americans; desegregation and the rise of the black athlete.

Not surprisingly, between the start of World War II and Billy Mills's triumph at the 1964 Olympics, few indigenous athletes attained national success. After the 1960s, an increasing emphasis on self-determination and a broader cultural rejuvenation led to the development of local and national athletic organizations that enhanced Native American athletic participation generally, and to a lesser extent in professional sports. Today, Native American athletes are less visible than they were a century ago; nonetheless, they continue to make noteworthy contributions to sports at all levels.

Race

Although not as critical to Euro-American understandings of Indianness as spirituality, warfare, or social organization, sports have contributed to efforts to make sense of, and correct, cultural differences. In fact, over the past two hundred years, most Euro-Americans have viewed indigenous athletes and athletics, like Native communities more broadly, in decidedly racist terms, stressing their primitiveness, physicality, and wildness. Racism and stereotypes, then, have been central to dominant interpretations of Native American athletes, no less than the opportunities granted them. Ward Churchill, Norbert Hill, and Mary Jo Barlow have argued that sports actually had negative implications for popular perceptions of Native Americans, particularly as they reconfigured existing stereotypes.

> The Native American within non-Indian mythology is (and has always been) an overwhelmingly physical creature. . . . [S]port was and is an expedient means of processing this physicality into a "socially acceptable" package without disrupting the mythology; Indians tracked as "Indians" into the mainstream. There could be but one result of such manipulation: dehumanization of the Native Americans directly involved and, by extension, dehumanization of the nonparticipating Native Americans whom the athletes represented in the public consciousness. Thus the myth of the American savage was updated but essentially unchanged. (Churchill, Hill, and Barlow)

The media have been central to the perpetuation of damaging racial stereotypes, including exaggerated attention to physicality, projections of savagery, the ever-present nickname "Chief," narratives centered on death, deprivation, and desolation, and clichés about drinking, excesses, laziness, lack of discipline, and an unwillingness to train. These images in turn shaped public perceptions. Moreover, fans have often shouted racial epithets, done war whoops, and engaged in other racist antics. At the same time, they increasingly incorporated Indian imagery into sports, choosing team names, logos, and mascots that played off of popular (mis)understandings of Indians and Indianness.

The history of indigenous peoples in sports demands comparison with African Americans. Native Americans never experienced outright segregation; however, they did experience overt discrimination, such as in rule differences in early lacrosse. Also, popular commentaries on the physical advantages and cultural deficiencies of Native American in the early twentieth century are remarkably similar to the ways in which African Americans are currently discussed in media accounts.

Indian Country

Throughout the twentieth century and to the present, indigenous athletes have had profound significance in Native American communities as well. In addition to the common pleasures and problems of sports spectatorship and participation, individual players, whether nationally celebrated or locally renowned, also have great social meaning in Indian country. Importantly, they have long facilitated the formulation of ethnic identity, tribal and pan-Indian; to take but one example, the Navajo golfer Notah Begay is a source of pride for both Navajos and Indians more generally.

Hidden from public view and too often outside of popular discourse, Native Americans have continued to reinvent sport and society. On the hand, they continue to play an array of traditional games, including lacrosse (Iroquois Nationals and Ojibwe Nationals), shinny leagues among the Tohono O'odham, and snowsnake among the Cree. On the other hand, they have incorporated Euro-American models and events, creating organization like Wings Across America and the Native American Sports Council, and establishing novel competitions like the Northern Games and the World Eskimo Olympics.

Athletes and athleticism have fostered efforts to improve indigenous lives and communities. Through his organization

Running Strong, Olympic great Billy Mills has endeavored to encourage character development and community outreach. More broadly, since the early 1970s the National Indian Activities Association has organized all-Indian competitions, promoting athleticism in harmony with indigenous ethics.

Others have seized upon sports as a means of ethnic revitalization. Members of the Iroquois nation in upstate New York, for instance, have established a national lacrosse team. In the process, they hoped to both enliven the rich heritage of ball play among the Iroquois and exert a claim that the Iroquois were a sovereign nation. The Iroquois National Team began participating in international competitions in the 1990s.

Terminology

Throughout this introduction, the terms *Native American, Indian, American Indian, Aboriginal indigenous,* and *indigenous peoples* have been used interchangeably to refer to the many diverse ethnic and political groups who originally inhabited the area now referred to as North America. The same strategy will prevail throughout the remainder of this work. This choice reflects both common usage and ongoing struggles over the politics of naming ethnic groups in the contemporary United States and Canada. Many find "Native American" to be preferable, because it parallels terms employed for other racial groups—for instance, African American—and avoids Columbus's misrecognition of the inhabitants of the Western Hemisphere as East Indians. At the same time, many reject "Native American." In addition to conservative commentators who think all who were born in the United States are native Americans, many of those to whom the name refers find it awkward, hollow, and overly academic; they think of themselves as "Indian" (or better, as members of a tribe) and use that to describe themselves and their peers. Adding in the Canadian context, in which "First Peoples" or "First Nations" are preferred, only complicates the situation. Also, several scholars and activists have recently proposed the term "indigenous" as more desirable. The editor's and contributors' use of all these terms interchangeably in part underscores the politics of naming without policing language or thought. It is hoped that refusing a single name will make clear to the reader complexities and diversities of living in and studying contemporary indigenous communities in North America.

Scope and Organization

Although often forgotten, and worse reduced to stereotypical symbols, Native Americans have had a profound impact on American sports. Indeed, indigenous athletes offer a powerful reminder that the history and significance of sport in American culture is incomplete without them. *Native Americans in Sports* seeks to present a thorough treatment of sport in Native America. It addresses athletes and athletics in all regions of the United States and Canada, offering an account of both the past and the present. It brings together biographies of athletes and profiles of all-Indian teams of national significance with overviews of specific sports, both indigenous and modern, to detail the Native American sporting experience.

Although every effort has been made to offer a comprehensive survey, the project has been hampered by the fragmented and incomplete nature of existing records of Native American athletes and athletics: some historic figures and games, known only superficially because of the paucity of sources, are treated only briefly, if at all; other figures long deemed Native had to be set aside because the evidence now indicates that they were in fact non-Native; still other figures championed by one or even several individuals could not be included because their achievements were too minor or regional. Despite these difficulties, this book does represent as complete a survey as possible of Native Americans and sports. It is hoped this work will serve not only as a reference, but will spark curiosity, dialogue, and inquiry, inspiring others to learn more and create a fuller, more dynamic understanding of indigenous athletes and athletics.

Clarence John "Taffy" ABEL

Born May 28, 1900, Sault Ste. Marie, Michigan
Died August 1, 1964, Sault Ste. Marie, Michigan
Hockey player

Clarence "Taffy" Abel (Ojibwe) first played organized amateur hockey with the Michigan Soo Nationals in 1918, in the U.S. Amateur Hockey Association, the first organized league in the United States. Abel competed for that team, also known as the Soo Indians, through the 1921–1922 season. In 1922, he joined the St. Paul Athletic Club of the Western Section of the new U.S. Amateur Hockey League. He played with them until he joined the U.S. Olympic team in 1924. At the first Olympic Winter Games in Chamonix in 1924, Taffy Abel became the first Native American to carry the U.S. flag in the opening ceremonies of an Olympic Winter Games, and he remains the only Native American to have carried the U.S. flag at an Olympic opening ceremony. The U.S. Olympic ice hockey team played five matches in 1924, winning four of them by double-digit margins but losing the final match to Canada, 6–1. Thus, Taffy Abel won a silver medal in his only Olympic appearance. He also scored fifteen goals in Chamonix, catching the attention of NHL general managers.

After the Olympics, Abel played one more year of amateur hockey with the St. Paul team and in 1925 joined the Minneapolis Millers. In 1926, Taffy Abel signed with the New York Rangers as a free agent. He played for eight years in the NHL. He was the first Native American to play in the NHL, and during most of his career he was the only Native American in the league. With the Rangers in 1927–1928, Abel became the first Native American Olympian to play on a Stanley Cup championship team. A six-foot one-inch, 225-pound defenseman, Abel was paired on defense with Ivan "Ching" Johnson on the Rangers. They were one of the toughest defenses in the league. In the Stanley Cup finals the Rangers' goalie, Lorne Chabot, was injured, and the Abel-Johnson defense supported backup goalie (and general manager) Lester Patrick well enough to enable the Rangers to win the cup.

Taffy Abel was traded to the Chicago Black Hawks after the 1928–1929 season. He played for five years in Chicago, retiring after the 1933–1934 season, but played on a second Stanley Cup championship team in 1934 with the Black Hawks. He retired at that point, after the Black Hawks' owner refused to give him a raise he had requested. During his NHL career, Abel had played 333 games and, in an era when defenseman rarely scored and no forward passing was allowed in the offensive zone, he totaled eighteen goals and eighteen assists. Huge for his era (he played at as much as 250 pounds) with a quick temper, he was a ferocious body checker and struck fear into most of his opponents. Taffy Abel was inducted as a charter member of the United States Hockey Hall of Fame in 1973. After his retirement from hockey, he operated Taffy's Lodge, a tourist resort in his hometown of Sault Ste. Marie.

Bill Mallon

Narciso Platero "Ciso" ABEYTA

Born December 15, 1918, Canoncito, New Mexico
Died June 22, 1998, Albuquerque, New Mexico
Runner, boxer

Abeyta, also named Ha So Da (Fiercely Ascending), lived on the Navajo Reservation as a child, attending the Santa Fe Indian School. He became student body president and caught the attention of Dorothy Dunn, a teacher who encouraged him to develop his artistic abilities.

During his high school years, Abeyta participated in track events, often running for miles at dawn, pushing his levels of endurance. He graduated from the Santa Fe Indian School in 1938. He was also a boxer, although records of his fights are difficult to find. During that era, Abeyta won the Golden Glove regionals and became New Mexico's state champion, participating in the national Golden Glove competition in Chicago.

He was also awarded an academic scholarship to Stanford University, but that opportunity did not come to fruition, as Abeyta instead served in the army during World War II, suffering from shell shock in its wake. Although it has been widely reported, even in his obituary, that Abeyta was a Navajo code talker, a stepdaughter says that this is incorrect.

After the war, he married Sylvia Ann Shipley, a potter, weaver, sculptor, and social activist. He also obtained a bachelor of fine arts degree from the University of New Mexico in 1952 and began working for the New Mexico Employment Commission.

Becoming more artistically prolific in the 1960s and 1970s, Abeyta received praise for his individualized style of Navajo art and his original ways of illustrating Navajo culture and creation mythology. He received many awards for his art, which appears in fine museums and university settings, including the Smithsonian Institution and the National Gallery of Art. His work also appears in exhibitions across the country and in Europe.

According to his stepdaughter, Alice Seely Warder, Abeyta continued to practice boxing techniques on a punching bag into his sixties, as well as jumping rope, and he enjoyed demonstrating his athletic abilities. "As children, we'd encourage him to walk on his hands," she said, "and then we'd collect the change that fell out of his pockets." Seely Warder also recalls three boxing trophies that decorated their home, including the Golden Glove award. "We played with them," she said, "but Dad was humble and down to earth and didn't fuss. Actually, we played with them until they broke."

At the age of seventy-nine, Abeyta died of a cerebral hemorrhage. Abeyta is buried in his family's cemetery in Canoncito, the place of his birth.

Kelly Boyer Sargent

FURTHER READING

http://users.1st.net/jimlane/98arch/6-26-98.htm.

ABORIGINAL SPORT CIRCLE

Established in 1995, the Aboriginal Sport Circle is the national body for Aboriginal sport in Canada. It is a collective of thirteen provincial and territorial Aboriginal sport bodies, each of which is mandated with responsibility for fostering sport and recreation opportunities for Aboriginal peoples within its region. The aim of this structure is to promote the development of the entire Aboriginal sport-delivery system in Canada, from the community through the regional and national levels of sport.

The ASC provides opportunities exclusively for the Aboriginal peoples of Canada as defined by the federal government through the Indian Act, including First Nations (status and nonstatus), Métis, and Inuit, from urban and reserve areas. While the ASC focuses on the development of a domestic, segregated sport system for Aboriginal peoples, it also encourages and assists Aboriginal athletes and coaches to take advantage of opportunities within the larger Canadian sport system. Toward this goal, the ASC focuses on mainstream sport activities.

The concept for the ASC emerged in the early 1990s, when a small group of Aboriginal sport leaders in Canada began lobbying the federal government, specifically Sport Canada, for financial support for Aboriginal sport. As part of their lobbying efforts, the Aboriginal sport leaders identified systemic barriers that severely limited Aboriginal athletes and coaches from participating in the Canadian sport system. As a result, the National Aboriginal Coaching and Leadership Program (NACLP) was established in 1993 to provide Aboriginal peoples with a national sport curriculum. Once the ASC was established in 1995, the NACLP was integrated into the larger organizational structure.

The national priorities of the ASC are athlete development, coaching development, and recognition of excellence. Athlete and coaching development are carried out through the annual National Aboriginal High Performance Training Camp, which encompasses an athlete training and coaching certification program. Annual since 1999, the camp has been held for basketball (1999 and 2000), volleyball (2001), and hockey (2001 and 2002) at locations throughout Canada. A new project is the National Aboriginal Hockey Championships. The inaugural championships was held jointly in the Mohawk territory of Akwesasne and Cornwall, Ontario, in April 2002. Both the camp and the championships celebrate Aboriginal cultural distinctiveness, while ensuring that Aboriginal athletes and coaches are provided with the necessary skills to integrate into the mainstream sport system should they wish to do so.

Recognizing the need to make coaching certification programs culturally relevant to Aboriginal people, the ASC developed a set of coaching manuals to be used in conjunction with the 3M National Coaching Certification Program, the national coaching certification program in Canada. The manuals deal with issues of racism in sport, traditional and holistic teachings, and health and nutrition for Aboriginal athletes. More than any other program area, the manuals demonstrate how Aboriginal peoples have developed and implemented innovative solutions to long-term systemic barriers in sport.

The ASC also annually recognizes the contributions of Aboriginal athletes and coaches in Canada through the Tom Longboat Awards and the National Aboriginal Coaching Awards.

Janice Forsyth

FURTHER READING

Aboriginal Sport Circle (www.aboriginalsportcircle.ca).

Paul ACOOSE

Born 1885, Sakimay Reserve
Died April 30, 1978, Sakimay Reserve
Runner

Paul Acoose was a long-distance runner whose brief career included victories in several races across Canada, both as an amateur and a professional, culminating in his defeat of the great Tom Longboat in a twelve-mile race in Toronto in 1910. He also set a world record for the indoor fifteen-mile event in Winnipeg on May 17, 1909, when he defeated English runner Fred Appleby with a time of 1:22:22. Wearing moccasins, Acoose showed an easy and regular stride with remarkable endurance, setting his pace early and outlasting his opponents in both short and long races.

Acoose, a Saulteaux, was born near Grenfell, Saskatchewan, in the spring of 1885. Both his father and grandfather were noted for their running and hunting ability. His grandfather originally had chosen the family name of Acoose, which means "flying bird" or "man above ground," as a testament to the running prowess of the family's men. Paul enjoyed distance running from an early age and ran in competitions before he was twenty. His early specialty was the five-mile event, which he won first in local events. By 1908, he was the Western Canada champion in both the three- and five-mile events. He was also victorious at the 1908 ten-mile Dominion Day race in Regina and the five-mile Labor Day race at Winnipeg. Seeking competition with better runners than were to be found locally, he turned professional in April 1909 and engaged a manager and two trainers.

In races in Winnipeg he defeated two well-known English runners, Alfred Shrubb and Fred Appleby. In a rematch against Appleby, Acoose was forced to drop out when a spectator spread tacks on the track, puncturing Acoose's moccasins. Undeterred, Acoose continued his success by winning a twelve-mile race against a two-man relay team in Victoria. By this time, Acoose was considered one of Canada's best runners. On March 12, 1910, at Madison Square Garden in New York, he placed second in a twenty-mile race against a top international field. His much-anticipated race against Tom Longboat, billed "The Redskin Running Championship of the World," finally took place on March 30, 1910, in Toronto. Acoose defeated the famous Onondagan runner in a twelve-mile race, Longboat dropping out after ten miles. This proved to be Acoose's last race. Homesick and weary of traveling, he returned home to the Sakimay Reserve, where he remained for the rest of his life.

He married in 1908 and later had nine children. In retirement, he farmed, tended cattle, and played an active role in the culture of the tribe as a respected Fancy and Grass dancer. Acoose continued to enjoy walking and running long distances, remaining active into his nineties. He died on April 30, 1978. He was inducted into the Saskatchewan Sports Hall of Fame in 1994.

Janice Forsyth

FURTHER READING

Zeman, Brenda. *To Run with Longboat: Twelve Stories of Indian Athletes in Canada,* edited by David Williams. Edmonton, SK: GMS Ventures, 1988.

Kenneth Jerry "Iceman" "Casper" ADAIR

Born December 17, 1936, Sand Springs, Oklahoma
Died May 31, 1987, Tulsa, Oklahoma
Baseball player

Kenneth Adair was a record-setting major league baseball player. He was a descendant of Cherokee tribal leaders from Adair County, Oklahoma.

Dubbed the "Iceman" because he was cool under pressure, Adair excelled at several sports in high school. During his senior year, he declined to be named as quarterback on the All-State football team, because it would make him ineligible to make the All-State basketball team. Later earning a spot on the All-State basketball team, he was the most outstanding player in the annual All-State game.

Adair was outstanding as a basketball and a baseball player at Oklahoma State

University during 1956–1958. He batted .454 his final season and made the All-Big Eight team and second team All-American. During the summer of 1958, he played for Williston in the Western Canada semipro league, where he led the loop in hitting at .409, tied for the lead in homers, and was the top-fielding shortstop. At the end of the Canadian season, the Baltimore Orioles signed Adair with a hefty bonus.

On September 2, 1958, Adair made his major league debut for the Orioles. He was with the Orioles until 1966. His teammate first baseman Jim Gentile nicknamed him "Casper," a reference to the cartoon character "Casper the Friendly Ghost," because of Adair's pale skin.

As an Oriole, Adair set two American League records for a second baseman—fewest errors in a season with 150 or more games (he had only five errors in 153 games in 1964) and most consecutive errorless games (eighty-nine games from July 22, 1964, to May 6, 1965). His glove is in the National Baseball Hall of Fame.

Adair's lifetime batting average for thirteen seasons ending in 1970 in the American League was .254. He also played for the Chicago White Sox, Boston Red Sox, and Kansas City Royals. He played second base for Boston in its loss to the St. Louis Cardinals in the 1967 World Series. After hitting .300 for Hankyu in Japan in 1971, he completed his major league baseball career as a coach for the Oakland Athletics and the California Angels. He had received three World Series rings as a coach for the series champion Oakland Athletics in 1972, 1973, and 1974.

Tragedy stalked Adair in his later years when he returned to his Sand Springs, Oklahoma, roots. Personal and financial problems forced him, always an introvert, into a shell. Kay, his wife and high school sweetheart, died of cancer in 1981. At the end of his playing career, their daughter, Tammy, one of four children, had died of cancer. One evening, though hoping to attend the thirtieth anniversary reunion of his college basketball team, Adair was admitted to a Tulsa hospital; he died the next morning of liver cancer. He was buried wearing his Baltimore Oriole uniform.

Royse Parr

FURTHER READING

Burke, Bob, Kenny A. Franks, and Royse Parr. *Glory Days of Summer: The History of Baseball in Oklahoma.* Oklahoma City: Oklahoma Heritage, 1999.

Hankins, Cecil. "Adair." In *Sand Springs, Oklahoma: A Community History.* Sand Springs, OK: Sand Springs Museum, 1994.

Kenneth Stanley "Bud" ADAMS, JR.

Born January 3, 1923
Football team owner

K.S. "Bud" Adams became a central figure in professional football as founder and longtime owner of the Houston Oilers/Tennessee Titans franchise. Throughout his life, Adams was also an amateur athlete, engineer, entrepreneur, businessman, and contributor to the Cherokee Nation.

Born Kenneth Stanley Adams, Jr., on January 3, 1923, Adams grew up in Bartlesville, Oklahoma. He took part in sports sponsored by his father, a former warehouse clerk who rose to become the chairman of Phillips 66, a multimillion-dollar petroleum company. Adams went on to play football and basketball at the Culver Military Academy in Indiana, Menlo College in California, and Kansas University, graduating with a bachelor of arts in petroleum engineering. During World War II, Adams trained at the University of Notre

Dame as an aeronautical engineer, then served a year in the Pacific theater in the U.S. Navy.

After the war, Adams moved to Houston, Texas, where he opened several Phillips 66 franchises and founded Adams Resources and Energy, Inc., to explore and drill for oil. Over subsequent decades, Adams built his fledgling enterprise into a Fortune 500 company. By 2002, Adams Resources and its subsidies formed a vertically integrated conglomerate that specialized in exploring for, producing, refining, and transporting oil, natural, gas, and petroleum, with Adams at the helm as CEO and president.

Adams's career in professional football began in 1959, when he paid $25,000 for the Houston Oilers, one of the eight teams in the newly formed American Football League (AFL). In 1966, Adams and the other AFL owners secured the future of the nascent league by merging with their rival, the more established NFL. The Oilers became a favorite team in Houston and throughout the state, despite its many failures to win an NFL championship, a frustration that often focused criticism on Adams. In 1997, after losing a battle with the city of Houston to have a taxpayer-financed football stadium built for the team, Adams relocated the Oilers to Nashville, Tennessee, and changed the team's name to the Tennessee Titans.

As a Cherokee Indian who traced his roots to ancestors in Tennessee before the era of Indian removal, Adams contributed time, money, and expertise to the Cherokee Nation. Adams served on the board of the Cherokee National Historical Society and the Cherokee Council of Economic Advisors, a group designed to help the tribe develop strategies to stimulate the local economy, create jobs, and offer assistance to individual Cherokee business owners. In 2000, Adams commissioned a series of sixteen oil paintings depicting Cherokee principle chiefs, donating it to the Cherokee National Museum. The same year, the Cherokee Heritage Society honored Adams and his service to the tribe by choosing him as featured guest and speaker at its annual fundraising dinner.

Nicolas Rosenthal

FURTHER READING

"Fortune 500: Adams Resources & Energy," *Fortune Magazine* (www.fortune.com/lists.F500/snap_17.html).

Fowler, Ed. *Loser Takes All: Bud Adams, Bad Football, & Big Business*. Atlanta: Longstreet, 1997.

Myers, Gary. "Oilers Hit Gusher in Nashville." *New York Daily News*, November 17, 1995, 79.

ADAPTATION

Many traditionally Native games have been adapted to fit mainstream "American" cultural standards, and many American sports have likewise been adapted to Native life. The ways in which those adaptations are contextualized are of utmost importance, though, for they shape perceptions of both mainstreamed and marginalized groups.

Lacrosse is perhaps the best example of a game adapted from a Native to a mainstream context. First recorded in the 1600s as a phenomenon among Indians in the Great Lakes area, the game seems to have existed in three basic forms (southeastern, Great Lakes, and Iroquoian). Different tribes played the game in different ways; for example, the southeastern tribes (such as the Cherokees) tended to use two sticks, cupping the ball between them, while the Great Lakes tribes (such as the Winnebago) used only one stick. Anthropologists quickly recognized and recorded Natives' passion for sport, and it is because of written accounts such as those by James Mooney in the late 1800s that the games' histories are known. In the case of lacrosse, that history reminds us that non-Natives did not take up the game until the middle of the nineteenth century. Now, the game has in excess of half a million players worldwide, though few people seem aware of its origins.

Those origins are much more complex than a simple translation of the game rules from one culture to another. As Thomas Vennum explains, Indian games are worlds of "spiritual belief and magic, where players sewed inchworms into the innards of lacrosse balls and medicine men gazed at miniature lacrosse sticks to predict future events. . . . [T]he magic is still there in the Indian game, but it is just under the surface, easily overlooked by the unsuspecting." Often, the "unsuspecting" are those who know of the game but not of its adaptation from Native to non-Native contexts. Also "unsuspecting" were those who wrote accounts that to their credit at least recorded the prominence of sport in Native culture. Joseph Oxendine laughingly notes one mid-nineteenth-century scholar who wrote, for example, that Indians paid too much attention to lacrosse and that they should "display the same attention to more important matters."

In spite of their obvious shortcomings, anthropological accounts do offer information on the existence of many games often thought to be traditionally Western. A sport similar to football, for example, is known to have existed among at least four tribes, and though little information about the game itself is available, the similarities are striking enough to point to likely adaptation of at least some basics into Western culture. Anthropological texts that chronicle Indian games are much more useful for tracing the adaptation of sports than they are for tracing the context in which those games were played. Accordingly, the context of the game often gets lost or redefined in the process of adaptation.

Until the late nineteenth century, however, Indian games remained largely confined to their traditional tribal contexts. But as the imposition of Western culture and values expanded and the theft of Indian land became more widespread, interchange of games among various cultures became more widespread as well. Absolutely crucial to the process of adaptation was the establishment of boarding schools, which mandated that Indian children be plucked from their tribes and actively assimilated into American culture. Part of that process of assimilation was an introduction to Western athletics. While at boarding school, many Indian students excelled in such sports as football and baseball, leading to a greater emphasis on individual athletes than had generally occurred in tribal contexts. Traditional Native emphasis on athletic endeavors in and of themselves shifted as a result of this interplay, so that the important part became less the sport itself than the person playing it. Texts such as *Indian Running: Native American History & Tradition* point to the ways in which this shift has altered the role of athletics in Native culture.

Adaptation of "American" sports to tribal contexts can sometimes be more difficult to analyze, largely because the games themselves remain mostly intact in terms of rules and appearance. For this reason, athletics and games cannot be divorced from their histories in Native cultures, for athletics have perhaps borne there a greater responsibility than entertainment or personal achievement. Mainstream basketball star Kareem Abdul-Jabbar certainly stumbled upon some of these differences while coaching a White Mountain Apache boys' team. He wrote of the difficulty of translating his own coaching style into a culture where players were uncomfortable when singled out. Similarly, observers often have a hard time avoiding the translation of long-standing, racist stereotypes of Indians themselves into conclusions about the athletes. Abdul-Jabbar made this tendency embarrassingly evident in his book, noting, "Sometimes I would glance his [the player's] way and imagine him sitting astride a paint pony two hundred years earlier, ready to ride off into the mountains and hunt." This kind of thinking hides the reality that Native athletes today are not products of "primitive" or overly masculine or romantic cultures any more than were their forebears. Rather, they are skilled sportsmen who have adapted nontraditional games into their necessarily multilayered existences. Basketball is but one of the sports adapted in this way. Although mainstream culture seems willing to embrace the image of an impoverished reservation dotted with scraggly basketball hoops and determined young people who have no other way out, it has been less willing to embrace more positive images.

Because athletics were used during the boarding school era as an assimilative tool for immigrant and minority populations, many people assume that the concept of adaptation is one-way, that although some Native games have evolved into modern-day games, Native populations have simply assimilated enough to learn to play mainstream sports. For this reason, there has been substantial resistance to acknowledgment that like mainstream cultures, Native cultures have adopted and adapted the sports to which they have been introduced, often reshaping the functions of the game, as in the oft-cited example of basketball.

Interestingly, though, in spite of the resistance to more complex analyses of such translations, mainstream cultures do not hesitate to use the image of the Native athlete as a symbol of its own values, such as good-spirited savagery. For this reason, Indian mascots are still prevalent in the sports world, and in spite of objections from many Native peoples that such images are degrading, mainstream cultures continue to enjoy the idea of "scalping" the competition or "tomahawking" their way to victory. Fans cheer on teams such as the Chiefs, the Braves, and the Indians, all the while resisting demands to situate such stereotypes in their historical contexts. When Native athletes first began to succeed on a large scale in mainstream sports such as baseball, other players would note the anomaly by labeling the player "Chief." Further, though running has a long tradition in Western as well as Native cultures, non-Natives now seem to want to connect their own histories to those of American Indians whenever the connection allows for some sort of New-Age-type mysticism. "It is to Indians that our imagination turns to brew imitative magic from tribal ritual," as Peter Nabokov explains it. Imitation remains at the heart of modern adaptation, certainly, but so too does exploitation, particularly of Native athletic ideals and individuals.

Adaptation, then, remains a complex concept, one that works not only in obvious ways such as the evolution of a particular game. It also emerges as a way to look at the translation of mainstream activities and values into nonmainstream contexts, where although appearances of the games are much the same, their social functions are entirely different.

Lisa R. Williams

FURTHER READING

Abdul-Jabbar, Kareem. *A Season on the Reservation: My Sojourn with the White Mountain Apaches*. New York: William Morrow, 2000.

Culin, Stewart. *Games of the North American Indians*. 2 vols. Lincoln: University of Nebraska Press, 1992.

Nabokov, Peter. *Indian Running: Native American History & Tradition*. Santa Fe, NM: Ancient City, 1981.
Oxendine, Joseph B. *American Indian Sports Heritage*. Champaign, IL: Human Kinetics, 1988.
Vennum, Thomas, Jr. *American Indian Lacrosse*. Washington, DC: Smithsonian Institution Press, 1994.

Amos AITSON

Born October 18, 1927, Carnegie, Oklahoma
Boxer

Aitson, a Kiowa, born to Franklin Aitson and Amy Geionety Aitson, became prominent in Oklahoma amateur boxing during the 1940s, taking the bantamweight title at the National Amateur Athletic Union boxing championships in 1945.

Aitson first became interested in boxing as a child, engaging in informal matches at a local grocery store. He later excelled in several sports, lettering in track and field for the Kiowa County Rural School in 1941, then lettering in both football and boxing in high school, the Riverside Indian School (Andanko [OK]), in 1943–1945. In 1944, he reached the state Golden Glove and AAU finals, finishing as runner-up both times to Bracey Murrow.

The following year, he was a member of a five-man Oklahoma City team that traveled to Boston for the fifty-seventh annual National AAU boxing championships in Boston Garden. A 118-pound bantamweight, Aitson reached the finals after defeating Charles Habron. On April 3, in a close fight, he decisively defeated Henry Levesque with a hard right to the jaw just as the final bell sounded. He also won the Oklahoma Golden Gloves bantamweight title in 1945, narrowly defeating Bud Love.

He enlisted in the navy in the summer of 1945 but was soon discharged owing to a minor physical defect. He returned to Carnegie but later moved to Wichita after getting married. He continued boxing for a time but also started wrestling professionally, traveling extensively with "Chief" Kit Fox. After a few years, however, he retired from sports, became a minister in the Pentecostal Church of God, and later built a church in Carnegie.

He moved to Dallas around 1960 and worked as a welder until retiring to Carnegie in 1991. In his retirement, he has served as director of the Kiowa Hall of Fame. He is married to Lucille Tsalote and has four sons and three daughters. He was elected to the American Indian Athletic Hall of Fame in 1982.

Edward W. Hathaway

FURTHER READING

"Boxing." In *The Britannica Book of the Year: Events of 1945*. Chicago: Encyclopedia Britannica, 1946.
Kidd, Wallace. "Franklin, Aitson Win." *Daily Oklahoman*, April 4, 1945.

Jackie Delane AKER

Born July 13, 1940, Tulane, California
Baseball player

Born in Tulare, California, a Potowatomi, Jackie Delane Aker began his professional baseball career in 1964 with the Kansas City Athletics, who signed him in 1958 to their minor league team. His parents—Cloud Thrasher Aker, a cotton farmer, and Lucille Trousdale, an amateur painter and Citizens Band Potowatomi—moved to California in 1938 from Oklahoma. Aker began playing Little League baseball when he was twelve years old. In high school Aker pitched and played football for Mount Whitney High School; he later pitched at the College of Sequoias in Visalia, California. In the minors he was an outfielder, but his hitting was so poor he went back to pitching.

Aker spent eleven seasons on seven different teams: the Kansas City Athletics (1964–1967), the Oakland Athletics (1968), the Seattle Pilots (1969), the New York Yankees (1969–1972), the Chicago Cubs (1972–1973), the Atlanta Braves (1974), and the New York Mets (1974), as a relief pitcher. In 1966 Aker tied the major league games-saved record of twenty-six and won the American League's Fireman of the Year award established by *Sporting News* and given to the top relief pitcher in each league. Aker played in 495 games, saved 123 games, won forty-seven games, and lost forty-five, compiling a lifetime 3.28 earned run average (ERA).

In 1967 the Athletics became entangled in a disagreement with owner Charles Finley, who suspended a pitcher for "rowdyism" and publicly admonished his players for a minor incident. Aker became the players' representative in the dispute, after which he played less and less. In all Aker played in 220 games, saving fifty-eight for the Athletics. While Aker never did as well as he did for the Athletics (back surgery in 1970 caused him pain throughout the rest of his career), he continued as a successful relief pitcher for both the Yankees and the Cubs.

When Aker retired from playing in 1974, he pursued managing in the minor leagues. A manager for ten years (1975–1984) he won the Manager of the Year award twice, the Governor's Cup award, and sent over ninety players to the major leagues. In 1985 he accepted a position as pitching coach for the Cleveland Indians, where he stayed until 1987.

In 1988 Aker established "Jack Aker Baseball," which offers camps, clinics, and individual instruction to children all over the country. Aker has also become involved with the National Indian Youth Leadership Project. Under a grant from the Project, Aker coaches Native American youth in Arizona and New Mexico. He travels to remote reservations and pueblos to teach baseball fundamentals and give talks stressing the importance of school and warning against drugs and alcohol. In 1997 President Bill Clinton presented Aker with a Giant Steps Award for his work with the National Indian Youth Leadership Project. Awarded every year by the Center for the Study of Sport and Society, the Giant Steps Award recognizes individuals for their dedication and work with student-athletes.

Lisa A. Ennis

FURTHER READING

Vecsey, George. "Sports of the Times: The Tutor from the Majors." *New York Times*, December 4, 1988, Late Edition, sec. 8, 3.

ALL INDIAN COMPETITIONS

North American Indians often gather to celebrate their culture through sport. Today, large-scale, international events, such as the World Eskimo-Indian Olympics and the North American Indigenous Games, and smaller scale multicommunity or intracommunity festivities, such as the Northern Games and Dene Games, continue to help North American Indians celebrate their connection to life on the land and keep their traditions alive.

World Eskimo-Indian Olympics

The first World Eskimo Olympics was held in Fairbanks, Alaska, in 1961. In 1973, this annual event's name changed to the World

An early example of all Indian competitions. The International Indian War Canoe Race, Coupeville, British Columbia, 1936. *(Seattle Post-Intelligencer Collection/Museum of History & Industry)*

Eskimo-Indian Olympics (WEIO) in order to reflect more accurately the ethnicity of Alaska. Like the Northern Games, the WEIO represents a celebration of the culture of the North, in particular Alaska. However, unlike the Northern Games, primarily a festival for participants and local spectators, the WEIO draws spectators from around the world.

The 2002 WEIO events included the Alaskan high kick, arm pull, blanket toss, dance team, drop the bomb, ear weight, Eskimo stick pull, fish cutting contest, four-man carry (no women), greased pole walk, Indian stick pull, kneel jump, knuckle hop, Miss World Eskimo-Indian Olympics contest, muktuk eating contest, Native baby contest (skin, cloth, fur), Native regalia, one-foot high kick, one-hand reach, race of the torch, seal cutting contest, scissor broad jump, toe kick, two-foot high kick, and white man vs. Native woman tug-o-war. Some categories for women were added in the 1970s, with the most recent addition being the 1983 addition of the sometimes gruesome knuckle hop.

The North American Indigenous Games

The first North American Indigenous Games (NAIG) were held in 1990 in Edmonton, Alberta, with three thousand athletes and cultural performers from Canada and the northwestern United States in attendance. Subsequent NAIGs were held in: Prince Albert, Saskatchewan (1993); Blaine, Minnesota (1995); Victoria, British Columbia (1997); and Winnipeg, Manitoba (2002). While the number of participants in each NAIG has varied, the 2002 Games had six thousand sport participants and three thousand cultural performers from eleven provinces/territories and sixteen states in the following sports: archery, athletics and cross country, badminton, baseball, basketball, boxing, canoeing, golf, lacrosse, rifle shooting, soccer, swimming, tae kwon do, volleyball, and wrestling. The NAIG Council, comprising thirteen members from Canada and thirteen members from the United States, plays a key role in creating the policies and procedures for NAIG while also overseeing the bid process to determine the host city for future NAIGs.

The Northern Games

The first Northern Games took place in Inuvik, Northwest Territories, Canada in July 1970. This Inuit sporting/cultural festival served three important roles. First, the games helped people to develop strength, endurance, and resistance to pain—all of which were important for survival. Second, the games helped to prepare them for life on the land. Finally, the games were used to celebrate culture. The Northern Games philosophy encourages "participation over excellence and an atmosphere of camaraderie and self-testing rather than competitive equality." Competition is not the driving force, and neither is time—the event schedule serves as only a rough guide.

After the initial Northern Games in Inuvik, subsequent games, which continued to involve participants from across Canada's territories and Alaska, were held on an annual basis for several years before moving to a regional format. Some of the more popular Northern Games activities include: one-foot high kick, two-foot high kick, Alaska high kick, one-hand reach, airplane, knuckle hop, knee jump, arm pull, musk-ox wrestling, head pull, good woman contest (includes duck plucking, tea boiling race, bannock making, seal skinning, fish cutting, and sewing competitions), drum dancing, fiddle and guitar, jigging and square dancing, blanket toss, the mouth reach, caribou leg skinning, muskrat skinning, bench reach, and wrist pull. While traditionally men and women did not compete in the same events, today it is common for women to participate in some male events, such as the kicking events.

The Dene Games

Dene games are played by the Dene, a group of people who live mostly in the Northwest Territories, located in Northwestern Canada. Dene games typically involve little or no equipment and were thus appropriate for the traditional Dene lifestyle, which involved traveling throughout the year. Heine has found that Dene games were often used as ways to practice skills that were needed for hunting and life on the land: strength, endurance, speed, and accuracy.

While Dene games used to be played on the land while traveling or during multi-community gatherings, Dene games have also been played at larger-scale festivals. The Dene-U Celebration Committee organized an event called "the Dene Games" in 1977, 1978, and 1979. A group called the Dene Games Association formed in 1980; the first Dene Games organized by this group took place in 1981 and continued on an annual basis until 1999. Due to the cost of travel and the cost associated with putting on such an event, multicommunity gatherings have become less frequent. Smaller intracommunity competitions often take place instead.

Popular Dene Games events include: archery, axe throwing, bannock making, bingo, canoe races, coin toss, Dene baseball, drum dancing, dry fish making, fish filleting and frying, hand games, log sawing, the spear throw, the stick pull, tea boiling races, and wood splitting.

Audrey R. Giles

See also: Compeiition Powwows.

FURTHER READING

Aboriginal Sport Circle of Canada (www.aboriginalsport circle.ca/naig.asp).
Heine, M. *Dene Games: A Culture and Resource Manual.* Yellowknife, NWT: Sport North Federation and MACA (GNWT), 1999.
Paraschak, V. "Discrepancies Between Government Programs and Community Practices: The Case of Recreation in the Northwest Territories." Ph.D. diss., University of Alberta, Canada, 1983.
———. "Variations in Race Relations: Sporting Events for Native Peoples in Canada." *Sociology of Sport Journal* 14 (1997): 1–21.
World Eskimo-Indian Olympics (www.weio.org).
Wulf, A. *Northern Games.* Inuvik, NWT: Author, 2002.

ALL INDIAN PROFESSIONAL RODEO COWBOYS ASSOCIATION

Following World War II, rodeos expanded and developed in the United States. The "big names" and fledgling sponsorships (equating with more attractive prize and place monies) continued to be associated with rodeos in Calgary, Pendleton, and Cheyenne, and Native-American performers had to find their places and create opportunities to demonstrate their talents. They demanded a rodeo administrative structure that would allow their rodeo achievements to be recognized at local, state, regional, national, and international levels.

By 1955 the open world of rodeo was managed by an organization called the Professional Rodeo Cowboys Association (PRCA). Its top event, known as the National Finals Rodeo, was seen to be the world championship. This administrative model was emulated by Native Americans, and in 1957 the All Indian Rodeo Cowboys Association (AIRCA) was founded. Key founding members were Dean Jackson, Jack Jackson, and Roy Spencer. The organization's name eventually changed to the All Indian Professional Rodeo Cowboys Association (AIPRCA). By the middle of the 1990s AIPRCA was drawing support from more than 600 members.

As various regional associations evolved under the AIPRCA umbrella, they created a competitive ladder that led to the institution of Indian national finals. Native American rodeos made sense financially *and* culturally. Regional or district rodeos reduced the travel costs for rodeo performers, and Native American rodeos took place as an important corollary to community celebrations, where the powwow was a vibrant barometer expressing tribal and ethnic continuity.

These associations emerged from, and were sustained by, Native American family groups, in which men and women shared the responsibilities. Accordingly, they created a dynamic entity that bore more resemblance to cultural avocations rather than recreational hobbies.

Calf roping at the Fort McDermitt Paiute Indian Reservation Rodeo. *(Library of Congress)*

Eventually AIPRCA came to comprise five regional associations in Canada and nine in the United States. These various associations are Indian Professional Rodeo Association (Alberta), the Indian Rodeo Cowboys Association (southern Alberta), the Western States Indian Rodeo Association (Washington and Oregon), the United Indian Rodeo Association (Montana), the Rocky Mountain Indian Rodeo Association (Wyoming and Idaho), the All Indian Rodeo Cowboy Association, the Southwest Indian Rodeo Association, the Navajo Nation Rodeo Cowboy Association, the Great Plains Indian Rodeo Association (North and South Dakota), the All Indian Rodeo Association of Oklahoma, the Western Indian Rodeo and Exhibition Association (British Columbia), the Northern Alberta Native cowboys Association, the Prairie Indian Rodeo Association (Saskatchewan), and the Eastern Indian Rodeo Association (Florida).

The focus for Native American rodeo was the Indian National Finals Rodeo (INFR) which was held for the first time in 1976, at Salt Lake City. From 1981 to 1994 the INFR was based at Tingley Coliseum in Albuquerque, New Mexico. In 1994 and 1995 the event took place at Rapid City, South Dakota. In recent years the INFR has leapfrogged from Saskatoon to Reno, Nevada, to Scottsdale, Arizona.

To the layperson the proliferation of regional associations and the shifting locus of operations seems to indicate fragmentation and discord within the AIPRCA; and factionalism and disagreements about representation and structure do exist. It is not surprising that in 1996 a new organization emerged, known as the All Indian Professional Rodeo Cowboys Association. While there was clearly a schism within Native American rodeo, it should be pointed out that stellar performers could and did compete in both INFR and AIPRCA events. "National" championships, for example, were not held on the same day.

Scott A.G.M. Crawford

FURTHER READING

Baillargeon, M., and L. Tepper *Legends of Our Times.* Seattle: University of Washington Press, 1998.

Iverson, P. *Riders of the West.* Seattle: University of Washington Press, 1999.

AMERICAN INDIAN ATHLETIC ASSOCIATION

The American Indian Athletic Association (AIAA) proved to be one of the most enduring of the many Native-oriented organizations designed to serve the burgeoning American Indian population of the Los Angeles metropolitan area. During the 1960s, the AIAA organized leagues for Indian amateur athletes in Los Angeles and held all-Indian tournaments that drew teams from throughout California and the American West. The AIAA's activities grew during the 1970s and continued through the 1980s and 1990s.

After World War II, American Indians joined the thousands migrating to Los Angeles, earning the city a reputation as the "urban Indian capital" of the United States and spawning dozens of organizations catering to the growing ethnic population's social, cultural, and recreational needs. In 1961, Indians in Los Angeles formed the city's first all-Indian sports association, the Inter-Tribal Indian Basketball League. Responding to a call for multiple sports, the group disbanded in 1964 and re-formed as the American Indian Athletic Association.

Basketball and softball became the most popular AIAA activities. League teams were organized or sponsored by other Los Angeles-area American Indian groups, such as the Los Angeles Indian Center, First Indian Baptist Church, Many Trails Indian Club, Little Big Horn Club, Indian Welcome House, Catholic Indian Club, Navaho Club of Los Angeles, and the Urban Indian Development Association. Separate divisions were established for men and women, and were open to anyone of at least one-fourth Indian ancestry. Additional groups spun off of the AIAA, such as the American Indian Bowling Association (AIBA), formed in 1966, and American Indian Youth for Sports, founded in 1971.

During the mid-1960s, the AIAA established annual tournaments that drew teams of Indian athletes from the Los Angeles metropolitan area and beyond. The Second Annual AIAA All-Indian Basketball Tournament, for example, held in Los Angeles in 1967, comprised thirty-four teams, including twenty from the city, fourteen from other parts of California, and eight from outside the state. Through the 1970s, the popularity of annual basketball and softball tournaments spread, and participation by teams from other western states continued to grow. Notably, these teams represented both rural and urban areas, making the tournaments an important arena for bringing together Indians of diverse backgrounds.

During the Reagan administration a drastic reduction in public funding for social services crippled many American Indian groups. The AIAA, in conjunction with the AIBA, persevered through participation fees, fund-raising, and the support of other Indian organizations. At the turn of the twenty-first century, annual Indian softball and bowling tournaments in the summer and Indian basketball tournaments in the winter remained highly anticipated events for hundreds of amateur Indian athletes and their families.

Nicolas Rosenthal

FURTHER READING

Bramstedt, Wayne Glenn. "Corporate Adaptations of Urban Migrants: American Indian Voluntary Associations in the Los Angeles Metropolitan Area." Ph.D. dissertation. University of California, Los Angeles, 1977.

Los Angeles American Indian Center. 1960s–1980s. *Talking Leaf.* Los Angeles.

Weibel-Orlando, Joan. *Indian Country, L.A.: Maintaining Ethnic Community in Complex Society.* Rev. ed. Urbana: University of Illinois Press, 1999.

AMERICAN INDIAN ATHLETIC HALL OF FAME

The American Indian Athletic Hall of Fame, a nonprofit organization, was incorporated in August 1972. It was, and continues to be, based at Haskell Indian Nations University in Lawrence, Kansas. From the start, it has sought to recognize great American Indian athletes. It was founded by Robert L. Bennett, former commissioner of Indian affairs, and Louis R. Bruce, his successor. The latter appointed Billy Mills, the 1964 Olympic gold medalist in the 10,000 meters, as the first hall of fame coordinator.

To be eligible for consideration as an inductee an athlete must be a member of a federally recognized tribe and be at least one-quarter American Indian or Native Alaskan blood or of sufficient degree to satisfy the Hall of Fame trustees. In addition, her or his sports participation (AAU, NCAA, or NAIA levels) must be of national prominence. Athletes, coaches, athletic directors, and other related candidates are evaluated for their athletic success, contributions to the Indian community, character, and success in chosen profession. Today, the candidate selections and inductions into the Hall of Fame are administered by seven trustees, the executive secretary, and the executive treasurer.

Eighty-four individuals have been inducted into the Hall of Fame in the past three decades. While football, baseball, and track and field dominate the sporting list, there are more unusual entries, including bowling, tennis, martial arts, swimming, and table tennis.

Scott A.G.M. Crawford

FURTHER READING

Oxendine, Joseph B. *American Indian Sports Heritage.* Lincoln: University of Nebraska Press, 1995.

Ross ANDERSON

Born May 8, 1971, Hollamon Air Force Base, Alamogordo, New Mexico
Speed skier

Anderson (Cheyenne-Arapaho/Mescalero Apache) describes himself as the "fastest Native American speed skier on Mother Earth." As late as the spring of 2002, Anderson was ranked ninth in the world and number one in the United States with a recorded speed of 146.694 mph in a sport that some have likened to various forms of auto racing without the cars. To gain some appreciation for the magnitude of Anderson's accomplishments, consider the observation of *Sports Illustrated* writer John Steinbreder that once speed skiers reach a thousand meters from the start, they are traveling at roughly 130 mph, "accelerating

Ross Anderson poses with trophies won on the 2000–2001 speed skiing circuit. *(AP/Wide World Photos)*

faster than a Ferrari." Given the effects of gravity forces on the body hurtling down the slope of a mountain, often at a pitch that would surpass the ability of most individuals to remain standing, the comparison to other "speed" sports is apt. Filmmaker Warren Miller included Anderson in his documentary feature *Freeriders*.

Born in New Mexico, Anderson was later adopted by a non-Indian family. He has spent most of his life in Colorado. Anderson grew up skiing, competing in the more familiar and traditional events of downhill and slalom. Demanding as those are, Anderson sought something even more "thrilling," gravitating toward a sport where aerodynamically designed race suits are the fashion of choice and speed is the name of the game. Sponsored by Running Strong for America's Youth, an organization founded by the great Lakota runner and Olympic gold medalist, Billy Mills, Anderson aspires to compete in the 2006 Olympics in Italy.

The fact that Anderson has achieved the success he has as an American Indian appears to be important to him. In one interview, Anderson explained that he had to search for his own identity, there being few role models portrayed in the mainstream press to whom he could look for guidance and support.

Responding to the need for American Indian role models, Anderson serves on the board of directors for the Native Voices Foundation, a nonprofit organization that sponsors programs designed to reach out to Native American youth by providing lift tickets, ski lessons, and equipment. In similar fashion, he has hosted the annual "Ski with Ross Anderson Weekend," an event intended to create opportunities for American Indian children to learn to ski, and spoken at events such as the Tenth Annual American Indian Science and Engineering Society (AISES) Regional Conference, encouraging American Indian children to pursue their dreams.

Ellen J. Staurowsky

FURTHER READING

Jackson, C. "Native Skier Has a Need for Speed." *Au-Authm Action News*, July 31, 2001, 25.

Ross Anderson biography (www.speedski.com).

Ross Anderson biography. *Running Strong for American Indian Youth* (www.indianyouth.org/sp_01_ross_Anderson.html).

Ross Anderson web site (www.rossanderson.org).

Steinbreder, J. "Faster than You Can Say Schuss." *Sports Illustrated*, January 27, 1992, 5–7.

AQUATIC COMPETITIONS

Aquatic activities were important to Native Americans. Many communities were located in proximity to the rivers, lakes, and oceans and so developed intimate relationships with them. They used waterways to move from one place to another, to bathe, and to nourish their bodies and souls. At the same time, they found rivers, lakes, and oceans sources of pleasure and amusement.

By all accounts, Native Americans were adept swimmers who enjoyed moving, playing, and being in the water. All ages participated in swimming. Children were taught to swim from an early age. Consequently, many displayed great skill and stamina. Some observers have even suggested that the crawl stroke began among the Mandan. Whatever the veracity of such assertions, Indian youth did display great inventiveness and imagination in the creation of games and contests. Swimming, however, never emerged as a native sport in its own right; it always remained instead spontaneous play and cherished recreation. In the 1950s, Robert Gawboy set a number of records in swimming.

Fishing was a common activity in many indigenous communities. In fact, for some native nations, it was fundamental both to their sustenance and their understandings of the world. Historically, fishing was never regarded as a sport by Native Americans. More recently, among the more important and intense political struggles for indigenous peoples in North America have been efforts to retain their right to fish in the face of encroachment by recreational fishing.

If fishing was a practical pursuit and swimming an amusement, boats allowed for the elaboration of a series of athletic competitions. This is not surprising given the importance of boats for transport and fishing, and in light of the fact that canoe and kayak are loanwords from indigenous languages. Contests that would be familiar to modern observers included races in which participants rowed for a set distance. Less familiar would be a variation on the tug-o-war in which two boats tied to one another would row in opposite directions until one had pulled the other a given distance. In some communities, a hybrid aquatic event was known; individuals would row to a marker and then swim back to the start. Native Americans have retained many of these events and participated in non-Native swimming events as well. Most notably was Canadian Jake Gaudaur, who three consecutive U.S. sculling titles in the 1880s and the world title in sculling in 1896.

C. Richard King

See also: Jacob Gill "Jake" Gaudaur; Robert Gawboy.

FURTHER READING

Oxendine, Joseph B. *American Indian Sports Heritage.* 2d ed. Lincoln: University of Nebraska Press, 1995.

Alexander ARCASA

Born 1890, Orient, Washington
Died 1962, Washington
Football player

Remembered by Glenn "Pop" Warner as "a young man of excellent character and disposition," Arcasa (Colville) excelled at Carlisle Indian Industrial School. Most noteworthy between 1909 and 1912 were his exploits in football and lacrosse.

Despite Arcasa's slight frame (standing five feet, eight inches and weighing 156 pounds), he was extremely versatile. Arcasa teamed in Carlisle's backfield with Jim Thorpe and Gus Welch, and he started games at quarterback and left halfback and right halfback. During his playing days at the school, Carlisle posted an impressive 39–11–2 record. Arcasa played on what was arguably Carlisle's best team: the 1912 squad. Carlisle won twelve games, lost one, and tied one. Walter Camp named Arcasa to his All-American team, and Arcasa's lacrosse teammates elected him captain of the best lacrosse team in the United States.

Arcasa actively participated in extracurricular activities while at Carlisle. In 1910, he read essays at special programs to honor guests at the school. After he graduated, Arcasa worked in railroad boiler shops in Altoona and later entered the U.S. Marine Corps, rising to the rank of colonel. After his death, he was elected to the Inland Empire Sports Hall of Fame in 1963 and the American Indian Athletic Hall of Fame in 1972.

William J. Bauer, Jr.

FURTHER READING

Newcombe, Jack. *The Best of the Athletic Boys: The White Man's Impact on Jim Thorpe*. New York: Doubleday, 1977.

Oxendine, Joseph B. *American Indian Sports Heritage*. Champaign, IL: Human Kinetics, 1988.

Steckbeck, John. *Fabulous Redmen: The Carlisle Indians and Their Famous Football Teams*. Harrisburg, PA: J. Horace McFarland, 1951.

ARCHERY

Indigenous peoples of the Americas have used projectile instruments for hunting, warfare, and sporting activities for tens of thousands of years. Early projectile weapons included hand-held spears, atlatls (or throwing sticks), and the darts they propelled. The bow and arrow followed, and by 500 C.E. all Native nations had the bow. Various bows were developed, reflecting differences among communities in their ecological settings and needs, cultural patterns, and social and ceremonial formation.

Types of Bows

Selfbows, made from staves of wood indigenous to whatever area the native nation occupied, were the most common. Local hardwoods were used, such as hickory, ash, black locust, white elm, ironwood, hornbeam, oak, osage orange (bois d'arc), and other woods, with yew, cedar, and spruce among the woods used in western coastal regions. There is little evidence of trade in bow stock. The limb cross-section was generally D-shaped, with the flat side of the D the back of the bow, the rounded part the belly. When strung, the limb curvature was distributed evenly throughout. Some selfbows were made with a reflex set, to give the arrow more thrust when shot. Flatbows, with wider limbs and more rectangular cross-sections, were also made in some communities. In the Great Lakes region, some selfbows used for ceremonial purposes had one or both sides carved in scallops. A sophisticated bow design was developed by the Penobscot people. This bow had a smaller set of limbs affixed to the back of the bow and attached by cordage to the longer set of limbs. Through the working combination of the double set of limbs, the design provided a mechanical leverage system, packing up to 30 percent more energy in proportion to the draw weight.

Another design, selfbows with static ears made by using heat to provide a set in the limbs at the tip ends, were used by the Woodland nations, and in certain other areas. This reflexed end, or static ear, does not bend but serves as a lever when the bow is drawn, concentrating more energy in the limbs, rendering the bow more powerful. Other selfbows, including short

George Catlin "Archery of the Mandans" depicts a traditional competition in the early nineteenth century. *(Denver Public Library/Western History Department)*

Plains bows as well as longer bows of the Iroquoian nations, were reflexed-deflexed, giving them a double-curved shape that was even more energy efficient. Such double-curved bows were sometimes backed with layers of finely shredded sinew soaked in glue. The combination of sinew-backing and wood core enhanced the bow's tensile strength while retaining its compression ability; the sinew-backing also added strength to the bow. Sinew-backed bows were found among the Plains nations and in areas extending from Athapascans as far south as the Apache, Navajo, and Pueblo homelands. Another very efficient type of bow was made from a combination of materials. Paiutes and others made wood composites, a combination of softwood for the belly, hardwood for the back, covered with a fine layer of sinew. Other composite bows, adapted for use on horseback, used a sophisticated combination of horn belly, wood core, and sinew-backing, creating a powerful bow with static ears. Antler or even buffalo rib was sometimes used in place of horn.

Certain Inuit bows used sinew to enhance performance, but instead of layering sinew directly on the bows Inuit bowyers fashioned the sinew into a cordage cable, which they then secured with an elaborate series of hitches to the back of the bow. Tension in the cables could be adjusted for weather conditions. Regional variations show up in the style of lashings, with sub-Arctic bows generally having fewer lashings than more northern bows.

Archery Games

The bow was part of the fabric of life, used in hunting, warfare, and ceremonial circumstances, as well as competitive games. Some involved stationary targets and tested accuracy at varying distances. Other games, using moving targets, demanded coordination of temporal and spatial trajectories, while yet others emphasized speed.

Ledger art drawings from the Plains nations provide documentation of some of these games. One piece done in the 1870s by a Cheyenne man, known as Making Medicine, and later in life as David Pendleton Oakhater, shows an archery gambling match—a combination of a social event, a competition between men's groups, and a chance for individuals to hone their reputations and skills. The drawing shows two competitors shooting at an upright arrow some distance away, as others watch. Two warrior societies have selected their best marksmen to compete. Each shoots an agreed-upon number of arrows toward the target, while another member of the society officiates, standing next to the upright arrow and shouting information about the shots to encourage or intimidate the participants. Whoever shoots the most arrows closest to the target wins.

Young Boys' Games

Young boys' archery games served to aid the development of spatial, temporal, and sensory motor skills needed later in life. One game played among the Lakota involved a small rolling hoop webbed with brain-tanned leather. Boys attempted to shoot arrows through the openings in the webbing as the hoop rolled rapidly over the ground, learning thereby how to orient themselves to a moving target. Another saw a boy wave a target made from the broad plate of a cactus plant, while others shot blunt arrows at it. Called *unkcélapte* in Lakota, the target was understood to represent the buffalo cow. The game recalled elements of a buffalo hunt, helping boys prepare for the time when they too would participate in hunts. Bows made from cherry branches, split in two, barkside on the bow backs, were used for this play; when the boys came of age they would begin to use the double-curved bows of adult men. Gender differences also influenced the use of the bow. Among some Native nations, women were not permitted to handle men's bows.

Speed Shooting

Accounts of speed shooting caught the attention of many nonnative observers, perhaps because prior to the introduction of the repeating rifle, well-skilled archers from the Plains nations had quicker firepower than gunmen. An often-repeated description of speed shooting appears in George Catlin's 1841 account of an archery game played among the Mandan. Holding a clutch of arrows in his bow hand, an archer would shoot his first arrow in a high arc to keep it in the air as long as possible. With quick movements he then discharged a second, a third, and so on in rapid succession. Whoever succeeded in keeping as many arrows simultaneously aloft won the game and whatever goods were staked on it.

This account, and others such as Longfellow's poetic visioning of Hiawatha's ability to keep ten arrows in the air, inspired a generation of non-Native sportsmen to take up speed shooting in the early twentieth century. Notable among this group was Saxton Pope (1875–1926), a medical doctor whose sports writings helped stimulate interest in archery. While a faculty member of the University of California, Pope worked with the Native

American known as Ishi (ca. 1860–1915). He was the last surviving member of the Yahi nation and had become a type of living exhibition in the university's Museum of Anthropology. By all accounts Ishi was a superb archer; photographs and descriptions attest to his skills. Pope learned much from Ishi, but like others of his generation, he believed European archery traditions superior.

As a North American sport, target archery dates to the nineteenth century. Notwithstanding the many different uses and types of bows found in Native America, the English longbow dominated the sport. Only since the 1970s have mainstream North American archers shown serious interest in Native archery traditions. Before that, the bows and arrows of Native America were seen as relics of vanishing ways of life.

Cherokee Cornstalk Shoot

Despite pressures to abandon their cultural identities, many Native nations maintained their traditions, including their competitive archery games. The Cherokee Nation's cornstalk shoot is one example. Highly social events, cornstalk shoots are part of the annual National Holiday celebrations commemorating the signing of the Cherokee Nation Constitution on September 6, 1839, and are held at other times. Eighty paces—about eighty yards, depending upon the stride of the man chosen to set it up—marks out the competition field. The targets are two ricks filled with cornstalks, set up at the opposite ends of the field. Participants stand by one rick, and each shoot two arrows at the other. Then they walk down to retrieve their arrows and shoot back up again, with one point given for each cornstalk penetrated by an arrow. A game might be called at fifty points, continuing for as long as it took someone to reach that score. Or it might be staged for a set period of time, with the winner the person with the most points when the time is up.

During the game, participants can joke and tease each other as they move between the two targets. At one time participants and villages bet against each other for food, money, and status. In the past men were the only competitors, but today cornstalk shoots can find women and children participating. Some cornstalk shoots allow the competitors to use only the more traditional selfbows, made of bois d'arc or other woods; others have different categories allowing the use of recurved and compound bows.

For other native nations, the importance of archery traditions diminished under the pressures of assimilationist policies. Reginald Laubin describes a citizenship ceremony that took place prior to the 1924 bill recognizing Native Americans as U.S. citizens. In this ceremony, a native man was handed a bow and arrow and instructed to shoot it far into the distance, suggesting the way of life he was leaving behind. Then, his hands were placed around the handles of a plow, representing his new road. Those pressures are now waning in face of current affirmations of native identity and cultural patterns. Among a number of native nations, bow traditions along with other traditions are being revived or made more public, and at places such as Rosebud, South Dakota, youth programs include the development of archery ranges.

Kay Koppedrayer

FURTHER READING

Afton, Jean, David Fridtjof Halaas, and Andrew E. Masich. *Cheyenne Dog Soldiers: A Ledgerbook History of Coups and Combat*. Denver: Colorado Historical Society and the University Press of Colorado, 1997.

Alley, Steve, and Jim Hamm. *Encyclopedia of Native American Bows, Arrows and Quivers*. 2 vols. New York: Lyons Press in Cooperation with Bois d'Arc Press, 1999, 2002.

Bailey, Judson. "Penobscot Bow." *Primitive Archer* 3:2 (1995): 6–11.
Catlin, George. *Letters and Notes on the Manners, Customs, and Condition of the North American Indians.* 1841. Reprint, Minneapolis: Ross and Haines, 1965.
Collins, Jeff. "Eskimo Bows, Part 1." *Primitive Archer* 6:1 (Spring 1988): 6–9.
———. "Eskimo Bows, Part 2." *Primitive Archer* 6:2 (Summer 1988): 5–7.
———. "Eskimo Bows, Part 3." *Primitive Archer* 6:3 (Fall 1988): 8–15.
Hamilton, T.M. *Native American Bows.* Special Publications, No. 5. Columbia: Missouri Archaeological Society, 1982.
Hamm, Jim. "Native American Archery Tackle." *Primitive Archer* 8:1 (2000): 18–26.
Herrin, Al. "Cornstalk Shooting." *Traditional Bowhunter* 8:3 (1996): 78–80.
Kroeber, Theodora. *Ishi in Two Worlds: A Biography of the Last Wild Indian in North America.* Berkeley: University of California Press, 1961.
Laubin, Reginald, and Gladys Laubin. *American Indian Archery.* Norman: University of Oklahoma Press, 1980.
Mason, Otis T. *North American Bows, Arrows and Quivers. Annual Report of the Smithsonian Institute 1893.* Washington, DC: Government Printing Office, 1894.
Oxendine, Joseph B. *American Indian Sports Heritage.* Champaign, IL: Human Kinetics Books, 1988.
Pinney, Dale. "Corn Stalk Shoot: How It's Done." *Primitive Archer* 6:2 (1998): 19–21.
Pope, Saxton. *Yahi Archery* 13:3 (1918).
Viola, Herman. *Warrior Artists: Historic Cheyenne and Kiowa Indian Ledger Art Drawn by Making Medicine and Zotom.* Washington, DC: National Geographic Society, 1998.

ARCTIC WINTER GAMES

The Arctic Winter Games (AWG) have been held biennially since their inception in 1970. The AWG were created to maximize participation opportunities in athletic opportunities for youth of all ages and cultures, or as many northern peoples' athletes as possible from all socioeconomic levels. The AWG seek to include as many athletes as possible in the games or the qualifying trials leading to the games. The AWG also provide an arena for northern sport competition for northern athletes who do not have the same competitive opportunities as athletes living south of the sixtieth parallel. Participation in the AWG provides northern peoples with an opportunity for social and cultural interaction.

The themes of athletic competition, cultural exhibition and social interchange that reflect the purpose of the games are represented symbolically by three interlocking rings and this symbol is the logo for the Arctic Winter Games International Committee.

The first AWG were held in 1970 in Yellowknife, Northwest Territories (NWT). A total of 710 athletes from the NWT, Yukon, and Alaska attended the first AWG. Since that time, the AWG have been held in: Whitehorse, Yukon (1972), Anchorage, Alaska (1974), Shefferville, Québec (1976), Hay River/Pine Point, NWT (1978), Whitehorse, Yukon (1980), Fairbanks, Alaska (1982), Yellowknife, NWT (1984), Whitehorse, Yukon (1986), Fairbanks, Alaska (1988), Yellowknife, NWT (1990), Whitehorse, Yukon (1992), Slave Lake, Alberta (1994), Chugiak and Eagle River, Alaska (1996), Yellowknife, NWT (1998), Haines Junction and Whitehorse, Yukon (2000), and Nuuk, Greenland and Iqaluit, Nunavut (2002). The 2004 AWG will be held in the Regional Municipality of Wood Buffalo (Fort McMurray), Alberta from February 29 to March 6, 2004.

Participation

Participation in the AWG has increased since its inception in 1970. The 1972 AWG included participants from the three originating bodies and also from northern Québec. Northern Québec also sent participants to the 1974 and 1976 AWG. Participation in the 1976 Games was reduced so as not to overburden the facilities available in Shefferville. Athletes from the original three locations attended the 1978 AWG. Seventy participants from northern Alberta and a small contingent from northern Québec joined athletes from the NWT, Yukon, and Alaska at the 1986 AWG. Fifty athletes and cultural performers from Greenland and cultural participants from the province of Magadan in northeastern Siberia joined the AWG in 1990. For the 1992 AWG, Greenland increased the number of athletes, coaches and mission staff to fifty and also sent ten cultural performers. Russia sent both cultural performers and athletes. The size of the team from northern Alberta had increased to two hundred participants. For the 1994 AWG, Alberta sent a full team. The 1994 AWG also included a seventy-member team from Greenland and two teams of thirty-five members from the provinces of Magadan and Tyumen, Russia. Over 1,600 participants, including athletes, coaches, mission staff, officials, and cultural performers, made the 1996 AWG a resounding success. The 2000 version of the AWG saw similar rates of participation. Team Chukotka (Russia) replaced Tyumen for the 2000 AWG.

The AWG include mainstream sports as well as traditional northern Aboriginal games developed by northern peoples in order to test and improve qualities such as strength, endurance, and balance. By 1996, the AWG included a total of nineteen sports. Sporting events in the 2000 AWG included: cross-country skiing, snowboarding, alpine skiing, hockey (men's and women's), badminton, basketball, curling, dog mushing, figure skating, gymnastics, indoor soccer, short track speedskating, ski biathlon, snowshoe biathlon, snowshoeing, volleyball, and wrestling. Traditional Inuit and Dene games (Arctic sports) are also included. A popular event is the one-foot and two-foot high kick. This game requires the competitor to jump off the floor and kick an object suspended overhead with either one or both feet, depending on the game played.

Unlike other major games such as the Olympics, no aggregate count is maintained of medal standings. The team that demonstrates the most sportsmanlike conduct throughout the AWG is awarded the Stuart M. Hodgson Trophy, named for the former commissioner of the NWT. This trophy is made from narwhal tusk and walrus ivory.

The budget for the first AWG in 1970 was four hundred thousand dollars. By 1990, the AWG budget had grown to $1.2 million; the cost of hosting the AWG is now roughly three times what is was in 1970. Support for the AWG comes from a variety of sources, including the governments of the Northwest Territories, Yukon, Alaska, Alberta, and Greenland. Various agencies and departments of the government of Canada also provide some assistance.

Susan Haslip and Victoria Edwards

See also: Gender Relations; Government Programs and Initiatives—Canada.

FURTHER READING

Alaska Native Knowledge Network. "Arctic Winter Games: A History" (www.ankn.uaf.edu/native games/winter-games.html).

"Arctic Sports a Tradition All Their Own." *ULU News* 7:1 (1982): 5.

Berrett, Tim. *Arctic Winter Games Economic Impact Statement: Final Report.* Rev vers. Submitted to the Arctic Winter Games International Committee, October 2000.

"History of the Arctic Winter Games" (www.awg.ca/History.htm).

Hurcomb, F. "Get Ready for the Winter Games." *Up Here: Life in Canada's North* 6:1 (Jan./Feb. 1990): 57–60.

Sports North of 60 Degrees. Fairbanks, AK: Fairbanks

North Star Borough School District, 1980, Videocassette.
"Whitehorse: Personal and Social Benefits of Participation" (www.awg.ca/Reports/AWG_2000_Social_Impact_Exeuctive_summary.pdf).
World Leisure Professional Services. *Executive Summary: 2000 Arctic Winter Games.* 2000 (www.awg.ca/Reports/AWG_2000_Social_Impact_Executive_Summary.PDF).

George E. ARMSTRONG

Born July 6, 1930, Skead, Ontario
Hockey player

George Armstrong captained the Toronto Maple Leafs when they won four Stanley Cups in the 1960s. He started his National Hockey League (NHL) career with the Leafs in 1949–1950 but spent the next two seasons with their farm team, the Pittsburgh Hornets, of the American Hockey League. He spent most of his youth near Sudbury. His father, of Irish descent, worked for the nickel mines in that city; his mother was Algonquin.

After some years in minor hockey in the Sudbury region, he was put on the Toronto protective list at age sixteen. He was quickly recognized for his offense skills. In 1947–1948, he led the Ontario Hockey Association (OHA) Junior A League in points, with seventy-three, while with Stratford, and moved to the Toronto Marlboros Junior the next season. In 1949–1950, with the Marlboros Senior, he helped his team to win the Allan Cup. He had 64 goals and 115 points that season. While playing for this series, the team had to play in Alberta, where the Stoney Reserve Indians called him "Big Chief Shoot the Puck." The nickname was with him for the rest of his life.

From 1952–1953 to 1970–1971, he played 1,185 games for the Toronto Maple Leafs, scoring 296 goals and 417 assists for 713 points. He also was penalized for 721 minutes. His best season offensively was in 1959–1960, with twenty-three goals and fifty-one points. During the winning playoffs of 1962, 1963, 1964, and 1967, he scored seventeen goals and made twenty assists, in forty-five games, to help his team win the coveted Stanley Cup. As he did before during the Allan Cup series, he rose to the occasion. In the history of the Maple Leafs, George Armstrong is fourth in points, first in games played, sixth in goals, fourth in assists, and he leads in most seasons spent with the team. He also won the J.P. Bickell Memorial Cup in 1959, an award given by the board of directors of Maple Leaf Gardens to the most valuable Leaf player. He is also, for the playoffs, the leader in games played and third in points. He played mostly at right wing but was asked many times to play center as well.

He became captain of the Leafs in 1957–1958, replacing Ted Kennedy who retired then. Armstrong was called by Conn Smythe, the owner of the team, "the best captain the Leafs have ever had." Smythe also honored him by naming one of his horses after him—Big Chief Army. Only two other hockey players were so honored by Smythe—Charlie Conacher and Jean Béliveau.

After his retirement in 1971, Armstrong became coach of the Toronto Marlboros, leading them to the Memorial Cup in 1973 and 1975, winning each time. He became a scout for the Québec Nordiques in 1978, before returning to the Maple Leafs as assistant general manager in 1988. But he was asked to go back behind the bench of the Leafs in 1988–1989 as an interim coach when John Brophy was fired. He is now a scout for the Toronto Maple Leafs.

The Toronto Maple Leafs won four Stanley Cups with George Armstrong as Captain. *(AP/Wide World Photos)*

He was inducted into the Hockey Hall of Fame in 1975.

Michel Vigneault

ASSIMILATION

"Assimilation" describes a set of policies and processes intent to acculturate Native Americans to mainstream white society and its values. Although it has its earliest expression soon after the founding of the republic, notably in the writings of Thomas Jefferson, assimilation did not crystallize as a popular ideology or focus of federal Indian policy until the late nineteenth century. Sport played an important role in efforts to assimilate Native Americans.

Once Native Americans were forced onto the reservation, the question arose of how to transform them into workers, citizens, and farmers. Beginning in roughly 1880 and countinuing through the 1930s, government programs and initiatives centered on assimilation. Native Americans were encouraged or forced to adapt Western dress, norms, values, language, religion and culture. Native Americans were forced to cut their hair, wear Western clothes, become farmers or ranchers, and adapt Western religions. Many Native American children were forced to go to one of the string of infamous Indian boarding schools, many miles from their families, to learn Western ways.

Throughout this period, sports played an increasingly important part of American society and were one of the chief ways of assimilating Native Americans into white American culture. The leaders of the assimilationist movement encouraged Native Americans on the reservations and the various boarding schools to play white sports. They hoped that participants would become fans of the various sports, or professionals, like immigrants from Eastern and Southern Europe. It was felt that if Native Americans became spectators or played the games, it would further their integration into white society. Consequently, Native Americans were steered toward the white sports of football, baseball, track and field, and basketball.

The Indian boarding school was one of the primary agents used to introduce Native Americans to white sports. Most boarding schools used sports as a part of the curriculum. A few boarding schools

Students at Carlisle Indian Industrial School play croquet. *(Library of Congress)*

achieved national recognition by using sports to assimilate Native Americans into American society. The Haskell Institute in Lawrence Kansas was one such institution. However, the most famous sporting Indian boarding school was the Carlisle (Pennsylvania) Indian School. From 1894 to 1914, the Carlisle fielded potent football squads that competed with the best and beat the most famous college teams of the era. The Carlisle School featured Jim Thorpe, one of the finest athletes who ever lived. Carlisle president Richard Henry Pratt argued that participating in white sports and beating the best white teams in the land showed Native Americans at the school, and back on the reservation, that they could compete with white men on the gridiron as well as in society.

Reservation Indians were encouraged to participate in the above-mentioned team sports, forming their own leagues on the reservations and sometimes competing against teams from other reservations.

The boarding school and assimilation experiment was a failure in its attempts at assimilating Native Americans into white society, but it did help introduce Native Americans to white sports and enable a few especially talented Native American athletes to escape the poverty of the reservation. However, many of the more famous Native American athletes discussed in this volume still fell victim to the traditional problems of their less athletically gifted brethren.

Rick Dyson

See also: Adaptation; Boarding Schools.

FURTHER READING

Adams, David Wallace. "More than a Game: the Carlisle Indians Take to the Gridiron." *Western Historical Quarterly* 32 (2001): 25–53.

Bloom, John. *To Show What an Indian Can Do: Sports at Native American Boarding Schools*. Minneapolis: University of Minnesota Press, 2000.

Fritz, Henry. *The Movement for Indian Assimilation*. Philadelphia: University of Pennsylvania Press, 1963.

Hoxie, Frederick E. *A Final Promise: The Campaign to Assimilate the Indians, 1880–1920*. Lincoln: University of Nebraska Press, 2001.

Michael Richard BALENTI

Born July 3, 1886
Died August 4, 1955
Baseball player, football player, football coach

Mike Balenti (Cheyenne) grew up on a farm in Darlington, Canadian County, Oklahoma. He learned to rope, ride, and handle stock at an early age. He prided himself as broncobuster and excelled at football and baseball.

Balenti entered Carlisle in September 1904. A teammate of Jim Thorpe, he quarterbacked the football team for several seasons. He signed a contract to play baseball with the Philadelphia Athletics, but following graduation from Carlisle in 1909 returned to football. Playing halfback at Texas A & M, he was named captain of the All-Southwestern team.

The same year, he began his professional baseball career, with El Reno in the Western Association, later playing with Dayton, Ohio (1909), Savannah, Georgia (1910), and Macon, Georgia (1911), before debuting in the major leagues with the Cincinnati Reds of the National League on July 19, 1911. The same year he married Ceili M. Barnovich (Haida).

He spent 1912 in Chattanooga, Tennessee. A year later he was on the roster of the St. Louis Browns and during the off-season was an assistant football coach at Saint Louis University. In 1914, he returned to Chattanooga and was named athletic director and football coach at the University of Tennessee at Chattanooga in July, while sitting out the end of the baseball season with a broken leg. Between 1915 and 1920 he was the assistant football coach at Baylor University. During this period, he played baseball for a number of minor league clubs, including San Antonio (1915), Galveston (1916–1917), and Tulsa (1917).

Balenti settled in Altus, Oklahoma in 1933, working in construction following his athletic career. He suffered a fatal heart attack in 1955 at age sixty-nine.

Richard Thompson

FURTHER READING

Special Collection. Baseball Hall of Fame. Cooperstown, NY.

BALL RACE

The ball race, sometimes referred to as a kick-ball race or relay, was a popular sport and significant cultural event practiced by indigenous peoples throughout the southwestern United States and Mexico, most notably among the Zuni, Pima, Tohono O'odam, and Tarahumare.

Ball races were traditionally conducted in the spring and might be played whenever the weather permitted. However, races were held more frequently after the planting season, when leisure time and favorable weather coincided. Both men and women participated, though never in the same race. Training for the ball race began in childhood, reportedly among children as young as four.

The race was a running contest in which participants continuously kicked a ball ahead of them along a prescribed course. Without breaking stride, runners used the tops of their toes to scoop up a ball, lifting it as they kicked and propelling it forward. In a typical kick, a ball flew approximately thirty feet and could roll an additional seventy feet or so while the runner caught up to it. Though a race might involve as few as two individual runners, it was typically contested by teams, each composed of four to six competitors. In team competition, all members of a team ran the entire course; one member of the team kicked the ball forward, and the others sprinted ahead. The team member closest to the ball when it landed would then kick the ball forward, and the process was repeated. Women played a modified version in which they used sticks to propel the ball or, in another variation, tossed a hoop.

Several types of balls were used for the race, the most common of which was not actually a ball but rather a cylindrical piece of wood or bone. According to contemporary accounts, these "balls" averaged three inches in length and one inch in diameter. Some tribes fashioned spherical balls, usually about 2.5 inches in diameter, out of stone or wood. Though not as common, balls constructed of animal bladders or skins were sometimes used as well. Because competitors normally wore sandals, which left the top of the foot unprotected, it was not uncommon for men to suffer extensive bruising after a long race.

The sport required a combination of speed, power, and skill. With each kick of the ball, runners sought to propel the ball as far as possible without sacrificing accuracy. Runners knew that a single misdirected kick could affect the outcome of the race, because the ball had to stay on the course; because they were prohibited from touching the ball with their hands, some carried sticks for use in recovering the ball in the event it strayed off course into a hole or other inaccessible location. Stamina was equally important, as races could cover distances up to twenty-five miles.

The courses, which were generally straight trails that led to a distant point and returned along the same route, were clearly marked. Judges were sometimes employed in static positions along the course to ensure a fair race. Spectators, many of whom wagered on the outcome, often followed behind the participants on horseback.

In preparation for a race, runners with long hair would tie it up on top of the head to prevent it from inhibiting their performance. The participants were usually attired in sandals and breechcloths and sometimes decorated their bodies and faces with bright paints. Major ball races were generally preceded by ceremonies that began the evening before the race and lasted throughout most of the night. Such ceremonies prepared the participants for the upcoming race through prayer, meditation, and ritual. The outcome of the race was seen as determined or influenced by supernatural forces; during such ceremonials groups looked for signs that might predict success or failure. The ball, in particular, was considered by most to possess a supernatural force that pulled the runners forward with it. The Pima of southern Arizona, in fact, reportedly believed that they could run faster while kicking a ball than they could run without it.

During the late nineteenth century, as a result of increased contact with Americans and under pressure from governmental and religious institutions to abandon traditional practices, southwestern tribes conducted ball races with growing infrequency. By the first two decades of the twentieth century, the ball race was no longer played in the United States except in boarding schools.

Caoimhín P. Ó Fearghail

See also: Kick Stick; Running.

FURTHER READING

Culin, Stewart. *Twenty-fourth Annual Report of the Bureau of American Ethnology, 1902–1903.* Washington, DC: Government Printing Office, 1907.

Hodge, F. Webb. "A Zuni Foot-Race." *American Anthropologist* 3 (1890): 227–31.

BASEBALL

Native Americans have contributed much to baseball, despite facing severe discrimination. Native Americans, unlike African Americans, were not excluded from organized baseball and therefore entered the game decades before them. However, they faced many obstacles, including racial stereotyping and degrading comments. The small number of Native American players and the reactions that they faced from fans and media helped to isolate them from other participants in the game. Despite these difficulties, Native Americans gradually increased in numbers, and many enjoyed considerable success.

Overcoming Stereotypes

An incident in the career of John McGraw, the Hall of Fame manager of the New York Giants, illustrates both the opportunities and difficulties that Native Americans faced during baseball's early history. Early in his managerial career, with the Baltimore Orioles in 1901, McGraw wanted to add a black player, Charlie Grant, to his team, but knowing that Grant would not be permitted to play if recognized as black, tried to pass him off as a Cherokee. McGraw, however, gave him the fictitious name Charlie Tokahoma—a play on "stroke a homer," reflecting the generally demeaning attitude that baseball usually expressed toward Native American culture.

Stereotyping of Native Americans was the rule in baseball as it was in other dimensions of American life in the late nineteenth and early twentieth centuries, including in the western dime novel and later Hollywood films. Native American ballplayers were regularly labeled as "Indian chiefs." The label was better than calling them "savages," as other entertainment media often did, but it demonstrated little cultural sensitivity. Among the so-called chiefs were Charles Albert "Chief" Bender, the Hall of Fame Chippewa pitcher for Connie Mack's Athletics; John "Chief" Meyers, a Cahuilla catcher for McGraw's Giants; George "Chief" Johnson, a Winnebago who pitched in both the National and Federal leagues; Allie "Big Chief" Reynolds, a Muscogee and clutch hurler with the Yankees in the 1940s and 1950s; and Louis Sockalexis, the enormously talented outfielder with the Cleveland Spiders in the late nineteenth century who was described as a descendant of Sitting Bull (although Sitting Bull was a Hunkpapa Sioux and Sockalexis grew up half a nation away as a Penobscot in Maine).

Bender may have been the greatest of all Native American players. Although a twenty-game winner only twice, he won 212 games in his career against just 127 defeats, three times leading the American League in won-lost percentage and compiling a lifetime 2.46 earned run average. Bender, considered by Connie Mack the greatest clutch pitcher he ever had, was inducted into the baseball Hall of Fame in 1953. Meyers, McGraw's regular catcher with the Giants for half a dozen years, batted .332, .358, and .312 from 1911 through 1913. His nine-year lifetime average was .291. Interviewed by Lawrence Ritter for Ritter's book on early stars, *The Glory of Their Times*, Meyers expressed great admiration for McGraw, who, despite his choice of a fictitious name for Grant, impressed Meyers as having an attitude toward both Native Americans and African Americans that was progressive for the times.

George Johnson won fourteen games with Cincinnati in his rookie 1913 season. He switched to the Federal League during the 1914 season, winning seventeen games for Kansas City in 1915, the final year of the league's existence. The New York Yankees acquired Allie Reynolds from Cleve-

An early Indian baseball team. *(Western History Collections, University of Oklahoma Libraries)*

land for the 1947 season, and for eight years he helped Charles "Casey" Stengel's Yankees continue their domination of baseball, winning between thirteen and twenty games each year as a Yankee. Reynolds's best season was 1952, when he won twenty, lost only eight, and led the American League in shutouts (six), strikeouts (160), and earned run average (2.06). During his tenure with the Yankees, the team won six pennants and six World Series championships, with Reynolds contributing seven Series wins against just two losses. An injured back forced his retirement after the 1954 season. Overall, Reynolds won 182 games and lost just 107.

Along with Bender's selection for the baseball Hall of Fame, three of the four players mentioned above (Bender, Myers, and Reynolds) were inducted into the American Indian Athletic Hall of Fame, an institution established in 1970 to honor Native Americans who excelled in one or more sports and who at the same time brought honor to themselves and to the Native American community. The Hall of Fame is housed in the Tom Stidham Student Union at Haskell Indian Nations University in Lawrence, Kansas.

Growing Impact on Baseball

Many other men of complete or partial Native American ancestry have played major league baseball. The most famous of them, although far from the best, was James Thorpe. Son of Sac and Fox parents, Thorpe was viewed in his time as the world's best athlete. Thorpe won college All-American honors twice in football

A baseball game in Macy, Nebraska. *(Library of Congress)*

while also starring not only in baseball but in archery, basketball, boxing, hockey, lacrosse, swimming, tennis, and track and field. He excelled in the 1912 Olympics as a member of the U.S. track and field contingent, winning both the pentathlon and decathlon. However, his prior experience as a semipro baseball player came to light in 1913, leading to confiscation of his gold medals. Thorpe then played professional baseball, primarily with the Giants, compiling a lifetime batting average of just .252. In a poll conducted in 1950, about four hundred sportswriters and broadcasters named Thorpe the best athlete of the first half of the twentieth century. He was named to the American Indian Athletic Hall of Fame in 1972.

The Johnson brothers, Robert Wallace and Roy Cleveland, had long and successful playing careers. Robert, like so many Native American players, endured a nickname linked to his heritage, in his case "Indian Bob." He played for Connie Mack's Philadelphia Athletics from 1933 to 1942 and completed his career with one season for the Washington Senators and two for the Boston Red Sox. An outfield regular throughout his career, he batted .296 with 2,051 hits. A productive power hitter, he hit over twenty home runs nine years in a row and batted in more than one hundred runs for seven consecutive seasons. His brother Roy, also an outfielder, played ten seasons, from 1929 to 1938 with Detroit, Boston, and New York in the American League and Boston in the National League. His lifetime batting average was identical to his brother's .296, but he lacked Robert's power, hitting only fifty-eight home runs in his career. However, he used his ability to make contact and run well to lead the American League in doubles in 1929 and triples in 1931. Austin Ben Tincup, a Cherokee from Oklahoma, experienced limited success as a pitcher, winning just eight games over a four-year major league career, all of them in 1914 with the Philadelphia Phillies. However, he enjoyed much more success as a long-time pitching coach and scout with the Browns, Phillies, Pirates, and Yankees. He was inducted into the American Indian Athletic Hall of Fame in 1981.

Gene Locklear, a Lumbee Indian from North Carolina, may have been the most artistic of all Native Americans in baseball. An outfielder, Locklear was a part-time player from 1973 to 1977 with Cincinnati and San Diego in the National League and the American League Yankees. He batted .321 with San Diego in 1975 but never quite fulfilled the promise he had demonstrated in the minors, where he won two batting championships and was twice named his league's most valuable player. His skill with an artist's brush was even greater, and longer lasting, than his artistry with a bat. One of his paintings, *The Tobacco Farm,* was selected to hang in the White House in Washington, D.C., and a print of the Dallas Cowboys was presented to President William Clinton.

The list of major league players of at least partial Native American descent, some of whom did not publicize their ancestry, includes pitcher Mose Yellowhorse, whose exploits in the early 1920s included hitting Tiger great Ty Cobb with a pitch after the batter yelled racial slurs at him; third baseman Pepper "The Wild Horse of the Osage" Martin, an important member of the St. Louis Cardinals' "Gashouse Gang" of the 1930s; pitcher John Whitehead, a productive starter for the Chicago White Sox for several years in the 1930s; first baseman Rudy York, a home run champion with the Detroit Tigers in the 1940s; and pitcher Calvin McLish, who won thirty-five games for the Cleveland Indians in 1958–1959; and many more.

Edward J. Rielly

See also: Guy Wilder Green; John Olson.

FURTHER READING

Fuller, Todd. *60 Feet Six Inches and Other Distances from Home: The (Baseball) Life of Mose YellowHorse.* Duluth, MN: Holy Cow!, 2002.
Oxendine, Joseph B. *American Indian Sports Heritage.* Champaign, IL: Human Kinetics Books, 1988.
Rielly, Edward J. *Baseball: An Encyclopedia of Popular Culture.* Santa Barbara, CA: ABC-CLIO, 2000.
Ritter, Lawrence. *The Glory of their Times: The Story of the Early Days of Baseball by the Men Who Played It.* New York: Vintage Books, 1985.
Wellman, Trina. *Louis Francis Sockalexis: The Life-Story of a Penobscot Indian.* Augusta, ME: Dept. of Indian Affairs, 1975.
Wheeler, Robert W. *Jim Thorpe: World's Greatest Athlete.* Rev ed. Norman: University of Oklahoma Press, 1979.

BASKETBALL

Basketball is immensely popular with Native Americans. It has arguably become the most important sport to indigenous people, who play and watch it with great passion. Although Native Americans have played a noticeable role in basketball within collegiate and professional arenas, the sport is much more important for most at the local level.

According to most accounts of the history of basketball, James Naismith invented the sport in Springfield, Massachusetts, in 1891. However, alternative versions, both intriguing and fantastic, of the sport's origins may be found in Native lore. In his magical prose, Sherman Alexie suggests that the sport actually was created by Aristotle Polatkin, who was shooting

Ute Girls play basketball in the Ignacio School gym, Southern Ute Agency, Colorado. *(Denver Public Library/ Western History Department)*

hoops before Naismith thought up the game. Perhaps more plausibly, others have pointed to two possible antecedents that may have informed the invention of basketball: hoop and pole, a sport in which competitors threw long sticks through a hoop; and pok-ta-pok, the forbearer of jai alai, in which two teams of five players each on a court of a hundred by fifty feet (basketball is played on a court ninety-four by fifty feet) tried to put a ball through a hoop mounted fifteen feet high (a basketball hoop is ten feet high) on the wall.

Whatever the precise origins of the sport, Native Americans have passionately embraced basketball, making it their own. Like many Western sports, indigenous peoples first encounter basketball in foreign contexts. Boarding schools introduced many Native Americans to the sport. Most institutions had teams by 1900, playing against white high schools and colleges in the West throughout the first quarter of the twentieth century. Other Native Americans first learned of the game from missionaries. In both settings, basketball, like sport more generally, was thought to be a fine way to assimilate indigenous peoples, teaching them core values of mainstream society. By the middle of the twentieth century, many reservation communities had established local teams and even hosted tournaments.

Native Americans have had a lasting impact on the sport of basketball. Over the past hundred years, a number of American Indians have excelled at both the amateur and collegiate levels. Historically, Stacy S. Howell, Clyde James, Jesse Rennick, and

Egbert Ward all gained wide notice for their play. More recently, a number of indigenous athletes have achieved fame in the National Basketball League, including Cherokee Parks, John Starks, and Kareem Abdul-Jabbar. Native women, although not as recognized as their male counterparts, have attained greatness as well. The Fort Shaw Indian School girls basketball team won the world championship at the 1904 World's Fair in St. Louis. Nearly a century later, female players like SuAnne Big Crow and Ryneldi Becenti continued to dazzle crowds, becoming legends for their prowess. Native Americans have not only played but coached basketball as well. Few have been as successful as Kelvin Sampson, twice recognized as coach of the year and currently head coach of the University of Oklahoma men's team.

Less visible to many in mainstream society are rich local traditions of basketball. In many communities, basketball has become as important as powwows. Few events attract as many people as a basketball game; passionate fans travel hours to watch them. Basketball tournaments stand at the heart of Indian basketball. On most weekends during the winter and spring, communities host intense competitions between teams throughout their region. Some tournaments have been contested for over fifty years. Teams travel great distances and pay hundreds of dollars for the chance to play others and take home pride, bragging rights, and prize money. Tournaments are not just for the young but typically include opportunities for young and old as well as men and women. These sports spectaculars often commemorate individuals.

The importance of basketball can also be read in the creative works of Native American artists and authors. Although the sport figures most prominently in the writings of Sherman Alexie, others, including James Welch, have turned to basketball to tell powerful stories about the contemporary Native American condition. In the film *Smoke Signals,* one of the protagonists, Victor Joseph, remarks, "Some days, it's a good day to die. Some days it's a good day to play basketball." Perhaps a greater testament to the significance of basketball in indigenous communities has been the recent spate of attention from mainstream society. *Chiefs,* a documentary by Daniel Junge and Donna Dewey, follows a high school basketball team on the Wind River Indian Reservation for two years. Larry Colton uses the life of a female basketball player to explore the complexities of American Indian life.

The future of Indian basketball looks bright. Undoubtedly in coming years, more Native Americans will play intercollegiate and professional basketball. The Native American Sports Council, in conjunction with the National Basketball Association, supports programs to enhance the skills of coaches and players. The Phoenix Suns sponsored the Native American Basketball Invitational in July 2003, a national tournament featuring the twenty-four best Indian high school teams (twelve teams of boys and twelve of girls). Others have talked of forming a professional Native American basketball league, composed of twelve to sixteen teams, each associated with a casino. In 2003, the Mohegan tribe purchased a WNBA franchise, the Connecticut Sun.

C. Richard King

FURTHER READING

Alexie, Sherman. "Why We Play Basketball." *College English* 58, no. 6 (1996): 709–13.

Allison, Maria T., Allison Lueschen, and Gunter Lueschen. "A Comparative Analysis of Navaho Indian and Anglo Basketball Sport Systems." *International Review of Sport Sociology* 14 (1979): 75–85.

Colton, Larry. *Counting Coup: A True Story of Basketball and Honor on the Little Bighorn.* New York: Warner, 2000.

Donahue, Peter. "New Warriors, New Legends: Basketball in three Native American Works of Fiction." *American Indian Culture and Research Journal* 21, no. 2 (1997): 43–60.

Robert Perry BEAVER

Born 1938
Football player and coach

Perry Beaver played professional football and coached high school football before becoming the principal chief of the Creek (Muscogee) Nation in January 1995.

In 1957, Perry Beaver graduated from Morris High School in northeast Oklahoma where he was an outstanding football center and linebacker. He was a junior college All-American at Murray Junior College in Tishomingo, Oklahoma. Thereafter, he played at Northeastern Louisiana University, where he signed as a free agent with the Green Bay Packers.

After being released by the Packers in summer drills, Perry signed with Grand Rapids, Michigan, in the United Football League, where he played for two seasons. He then returned to Oklahoma to work on his degree at Oklahoma Central State University, graduating in the summer of 1966. That fall he became an assistant football coach at Jenks High School, a large public school in a suburb of Tulsa, Oklahoma. Beaver was named the head football coach at Jenks in 1977 and led his teams to state championships in 1979 and 1982. He also served as a teacher and director of Indian education in the Jenks public school system. His composite football coaching record at Jenks was 109 won and 53 lost.

In 1991, his peers elected him to the Oklahoma High School Coaches Hall of Fame. At the time of his hall of fame award, Beaver announced that he would run for principal chief of the Creek Nation. He had been elected the Creek's second chief in 1971 and had served on the Creek Council. Beaver lost his first election for principal chief in December 1990 but was elected to four-year-terms as principal chief for terms beginning in January 1995 and January 1999.

After his first election as principal chief, Beaver stated that running an Indian tribe would be similar to coaching a football team, in that it would involve everyone working together. Beaver stressed that treaties gave the federal government a responsibility to provide tribal health, education, and welfare programs. He promised to be a champion of Indian sovereignty and to give education and job skills priority in his administration.

In 2000, Beaver was inducted into the American Indian Athletic Hall of Fame.

Royse Parr

FURTHER READING

Hibdon, Glenn. "Jenks Produced Big Linemen During Beaver Coaching Era." *Tulsa World*, July 28, 1991.

Martindale, Rob. "Teamwork Is the Game Plan for New Creek Nation Chief." *Tulsa World*, January 7, 1996.

Ryneldi BECENTI

Born 1971
Basketball player

At age twenty-one, Ryneldi Becenti (Navajo) was called "the most accomplished athlete of her generation." She was the first Navajo to play college basketball at a major institution, breaking records during her years at Arizona State University, and playing professionally after graduating.

Becenti, called "Sis" by her family, was born in 1971, the middle child out of five children and the only girl. Her grandparents and parents played basketball. Her mother Eleanor—who passed away when Ryneldi was a teenager—played in Indian tournaments while she was pregnant with Ryneldi. Additionally, her brothers played the game, though none with the passion and persistence of their sister. Throughout her career, Becenti practiced her game constantly.

Becenti joined the Navajo Nation's team, and as a high school sophomore she began playing on the Window Rock High School team. The next year her team won the state 3A championship. As a senior in 1989, she was named Arizona's high school Player of the Year. She moved on to Scottsdale Community College, where she set records for that school, received more titles, and earned the attention of coaches at Arizona State University, which offered her a scholarship.

At five feet, seven inches, Becenti relied more on agility and speed than height. During her time as the starting point guard for ASU, Becenti led the conference in steals and assists. She was also a role model and inspiration for other Navajo, who began to play college basketball in greater numbers. A two-time All-Pac Ten guard, she was an honorable mention All-American in 1993.

Becenti also holds ASU's record for assists in a single game, with seventeen. Her ASU career average of 7.1 assists per game remains the best in the Pac Ten. In July 1993 Becenti went to the World University Games, the only Native American out of 362 athletes representing the United States.

In 1995, Becenti fulfilled a personal dream and became the first Native American to play basketball with a European team. She played in Switzerland, a year later returning to the WNBA and the Phoenix Mercury. In 1996 she was inducted into the American Indian Athletic Hall of Fame, the second Navajo and the first Navajo woman to be so honored.

Because of basketball, Ryneldi Becenti has long been a role model, talking to groups on and off the reservation. "You could hear a pin drop when she spoke to kids," says Peterson Zah, past president of the Navajo Nation. Her coaches also stressed her importance as a Native American. "She's a very, very big deal to us Navajos," said Len Kinsel, a junior high school athletic director. According to Maura McHugh, Becenti's head coach at ASU, "She is more than just a basketball player to a whole lot of people."

Since earning her degree in sociology from Arizona State University, Becenti has lent her support to community groups and special projects. She has led workshops for youth and spoken at tribal conferences. She has joined all-Indian teams to play in tournaments for the National Indian Athletic Association.

"My main objective," Becenti said in 1993, "is to return to the reservation and give what I've experienced and what I've learned to younger kids, so they can go out and explore the world.... [W]hen I go back and tell them, it gives them a little more motivation to leave the reservation and be a hero, too."

Vickey Kalambakal

FURTHER READING

Aaseng, Nathan. *Athletes*. New York: Facts on File Books, 1995.

Rubin, Paul. "Arizona's Shooting Star." *Phoenix New Times*, 23:51: 26–33.

Cheri BECERRA

Born 1980, Nebraska City, Nebraska
Wheelchair racer

Born in Nebraska City, Nebraska, and a member of the Omaha tribe, Cheri Becerra by her athletic accomplishments and motivational speeches has positioned herself as an incredibly positive role model. At the age of four Becerra contracted a virus that paralyzed her from the waist down. Despite her paralysis Becerra was a very active child and young adult. Her love of swimming as a child helped to develop upper-body strength that propelled Becerra to a Paralympic champion in wheelchair racing.

Although Cheri Becerra did not begin racing until she was eighteen years old, she quickly moved into the upper competitive circles of women's wheelchair racing. At a Junior regional meet in Wichita, Kansas, in 1994 she sat in her first racing wheelchair and won events in that chair the very same day.

Most in the sport, however, thought it impossible for the newcomer Becerra to prepare for the 1996 Paralympics in Atlanta. With the financial assistance of several local, regional, and national Native American organizations Becerra was able to defray the costs of travel and equipment as she prepared for the 1996 games. Silencing her critics, Becerra not only qualified for the 1996 Paralympics but brought home medals in several events. She won silver in the 100 meters, with a time of 16.74 sec-

Cheri Becerra wins the 100 Meters at 2000 Paralympics, Sydney Australia. *(AP/Wide World Photos)*

onds and the 200 meters in 29.64 seconds. Becerra's third-place finishes, in the 400 meters and the 800 meters, earned bronze medals. The 2000 Paralympics in Sydney, Australia, were just as successful for Becerra. Finishing just over one second from the leader in the 800 meters, Becerra finished fifth in the event. She took gold medal honors in the 100 meters and the 400 meters. In addition, Becerra set a world record in the semifinals of the 200 meters in 28.78, the only person to break twenty-nine seconds in that event. Beyond her achievements in the Paralympics, Becerra has had success in ten-kilometer events. She is a two-time winner (1999 and 2000) of the Bolder Boulder (CO) 10K road race.

Becerra's accomplishments on the track are made even more remarkable by her straightforward, no-nonsense approach to training. She defies conventional wisdom by training by herself, for the most part without the guidance of a coach. Observers credit Becerra's success to an incredible work ethic and confidence. Becerra also credits the support she has been given by her family.

Through the Native American Sports Council, Cheri Becerra commits some of her energy to giving motivational speeches within Indian communities and reservations around the nation. Her determined and optimistic spirit inspires those who hear her words. For that and her athletic achievements Becerra was awarded the Nebraska Statewide Citizens Award in 1996 and the City of Lincoln Spirit award a year later.

Rita M. Liberti

FURTHER READING

"Cheri Becerra Story." *NdnSports.com* (www.ndnsports.com/columns).

Gallo, Jim. "People in Sports: Cheri Becerra, a Family Affair." *Sport 'N Spokes* 23:7 (1997): 40–44.

W. George BEERS

Born May 5, 1843, Montreal
Died December 26, 1900
Lacrosse player

William George Beers, more than any other individual, promoted the sport of lacrosse as Canada's national game. As a player, rule maker, author, administrator, and international tour organizer, Beers fervently promoted lacrosse as an exercise in patriotic manhood. Beers was responsible for the longstanding myth that lacrosse was decreed the national sport by Canadian parliament during the immediate post-Confederation period. In effect, this lobbying and Beers's influence had a significant impact in the official naming of lacrosse as Canada's national summer sport more than one hundred years after his original claim.

W. George Beers was born in Montreal, and played his first game of lacrosse at age six. He was educated at Philips School in Montreal and Lower Canada College before training to become a dentist. Beers's lacrosse abilities gave him the opportunity to play goal at age seventeen in a lacrosse match staged for the prince of Wales during his tour of Canada in 1860. The prince was particularly interested in matches that featured white lacrosse club teams against Native Canadians. He was also supportive of the promotion of the sport of lacrosse to other parts of Canada. Beers, working as a dentist and publishing the *Canadian Journal of Dental Science*, became a member of the Montreal Lacrosse Club (MLC) and secretary of the newly formed National Lacrosse Association, formed in 1867.

Beers immediately began to promote the sport of lacrosse as Canada's game, a test of manhood, virility, and one of the "true" Canadian sports. All other sports, he claimed, were foreign. Beers published *Lacrosse: The National Game* in 1869, a book that outlined the rules, positions, strategies, and techniques of competitive lacrosse, and he subsequently wrote widely on the merits of the sport in various magazines and newspapers. Beers raised his arguments about the strength of Canadian manhood to both the Americans and the British. He organized a tour of Scotland, Ireland, and England in 1876 to promote the sport and its celebration of Canadian sporting manhood. In 1883 he helped to arrange a tour during which the teams distributed a variety of information and gave lectures to the spectators to promote immigration to Canada.

Beers wrote widely on the Native origins of lacrosse to substantiate his claims that lacrosse was distinctly Canadian. Of course, many Native groups residing in United States territories had played the same type of game. His patriotic campaign focused on having lacrosse declared as Canada's national sport, over cricket and other "foreign" exercises. So vociferous and prolific on this matter was Beers that a longstanding myth emerged that Canadian parliament during the late 1860s had named lacrosse as the official national sport. There is no record of such an act. However, in 1994 Canadian parliament declared lacrosse to be Canada's official national summer sport. No other individual promoted the sport of lacrosse in Canada to the extent that W. George Beers did. Indeed, his writings and early influence played a major role in its official declaration as the national game. Beers died in 1900.

Kevin B. Wamsley

See also: Lacrosse.

FURTHER READING

Beers, G.W. *Lacrosse: The National Game of Canada.* Montreal: Dawson Bros., 1869.

Lindsay, Peter L. "George Beers and the National Game Concept: A Behavioural Approach." *Proceedings of the Second Canadian Symposium on the History of Sport and Physical Education.* Windsor, Ontario: University of Windsor (May 1–3 1972): 27–44. Ottawa: Sport Canada Directorate, 1972.

Morrow, Don. "Lacrosse as the National Game." In *A Concise History of Sport in Canada,* edited by Don Morrow et al. Toronto: Oxford University Press, 1989.

Notah Ryan BEGAY III

Born September 14, 1972
Golfer

Born to a Navajo father and a Pueblo mother (San Felipe and Isleta descent), he is the only full-blooded Native American currently on the Professional Golf Association Tour. He spent much of his youth on a reservation in New Mexico, moving to the state capital only after scoring well enough on an entrance exam to attend Albuquerque Academy. Unlike many golfers, Begay overcame the underdevelopment of the reservation and his family's lower-middle-class position to excel as an athlete and role model.

Begay began playing golf at an early age, perhaps as young as six. He refined his game on municipal courses. Throughout his youth, he displayed passion for golf, sneaking onto courses to play and later working, not for money, but for playing time. He won more than a dozen junior and amateur titles during his teens. Importantly, Begay was not merely a golfer. He played soccer and basketball in high school, leading his high school to state championships in the latter sport in 1989 and 1990. He was named the New Mexico Athlete of the Year in 1990.

After graduating high school, Begay attended Stanford University, earning a degree in economics. During his four years, his teammates included Tiger Woods and Casey Martin. Together they won the na-

Notah Begay, the only full-blooded Indian on the PGA Tour, has won four major tournament events. *(AP/Wide World Photos)*

tional championship in 1994. Begay earned All-American honors three times, and he set the record for a single round in the NCAA Championship, shooting sixty-two.

In 1995, Begay turned professional. After three years, he began to hit his stride on the Nike/Buy.com tour. He was runner-up on four separate occasions. In 1999, he joined the PGA tour. In his first two years he won four events: the Reno-Tahoe Open (1999), the Michelob Championship at Kingsmill (1999), the Fedex St. Jude Classic (2000), and the Canon Greater Hartford Open (2000). Due in part to his two majors victories in 1999, he was nominated for Rookie of the Year honors and holds a lucrative endorsement contract with Nike.

In spite of his excellence and exploits, many know Begay only because he was arrested for drunk driving in January 2003. In a move that surprised many and heartened others, he took responsibility for his actions and even informed prosecutors of a previous conviction in Arizona five years earlier. Begay was given a 364-day sentence with all but seven days suspended, a thousand-dollar fine, and forty-eight hours of community service, and was allowed no alcohol consumption for a year. Although some have criticized the terms of plea—namely, the work-release during his brief incarceration that enabled him to spend up to half of the day outside of prison, training and playing golf—as special treatment,

Begay has retained endorsement contracts and arguably polished the luster of his celebrity.

C. Richard King

FURTHER READING

King, C. Richard. "A Notable Exception: Notes on Notah Begay, Race, and Sports." In *Telling Achievements: Native Americans Athletes in Sport and Society*, edited by C. Richard King. Lincoln: University of Nebraska Press, forthcoming.

Johnny Lee BENCH

Born December 7, 1947, Oklahoma City
Baseball player

Johnny Bench is considered by many to be among the best catchers to ever play the game. *(AP/Wide World Photos)*

Bench ranked among the greatest major league baseball catchers and made major league baseball's All-Century team. Bench, one-eighth Choctaw Indian, was the son of Ted Bench, a truck driver and semiprofessional baseball player, and Katie Bench. Bench earned All-State honors in baseball and basketball at Binger High School, graduating as class valedictorian in 1965.

The Cincinnati Reds selected Bench in the second round of the 1965 amateur draft. Bench rejected several college scholarship offers and signed a bonus contract with the Cincinnati Reds. He won Carolina League Player of the Year honors with Peninsula in 1966 and garnered the *Sporting News* Minor League Player of the Year honors with Buffalo of the International League in 1967.

The much-heralded Bench joined the Cincinnati Reds at midseason in 1967. A year later, he set National League records for most doubles by a catcher (forty) and most games caught by a rookie (154), and became the first catcher named National League Rookie of the Year.

The six-foot one-inch, 208-pound Bench demonstrated durability, power, and a cannon-sized arm, bewildering base runners with laser-like throws to second base. He also possessed huge hands and catlike quickness; he redefined the position with his one-handed catching style and the left-handed sweep tag. The innovative Bench introduced the helmet and the oversized hinged mitt that became part of every catcher's equipment.

Bench batted either fourth or fifth for the "Big Red Machine," which won six National League Western Division titles, four National League pennants, and two World Series in the 1970s. He earned the Most Valuable Player award in 1970 and 1972 and the World Series Most Valuable Player in 1976. He earned ten consecutive Gold Gloves, from 1968 through 1977, for outstanding defensive play and was named to fourteen National League All-Star teams. Bench compiled a .267 career batting average, hit 389 home runs, and drove in 1,376 runs. His figure of 327 career home runs ranks second among major league catchers.

Bench married Vicki Chesser, a high-profile New York model. The marriage ended in divorce. Bench hosted television and radio shows, including *MVP: Johnny Bench* from 1971 to 1976 and *The Baseball Bunch* from 1981 to 1983. Bench joined CBS Sports as a baseball commentator in 1985 and was elected to the National Baseball Hall of Fame in 1989. In 1999, he rejoined the Cincinnati Reds as special consultant to the general manager.

David Porter

FURTHER READING

Bench, Johnny, and William Brashler. *Catch You Later: The Autobiography of Johnny Bench*. New York: Harper, 1979.

Vecsey, George. "Johnny Bench: The Man Behind the Mask." *Sport 54* (October 1972): 101–12.

Walker, Robert H. *Cincinnati and the Big Red Machine*. Bloomington: Indiana University Press, 1988.

Hall of Fame pitcher, Charles Bender, like many other indigenous athletes, excelled on the field in spite of racism. *(Library of Congress)*

Charles Albert "Chief" BENDER

Born May 5, 1884, Brainerd, Minnesota
Died 1954
Baseball player

Bender helped pitch the Philadelphia Athletics to five American League pennants and three World Series titles between 1905 and 1914. He was one of thirteen children of Albertus Bliss Bender, a farmer of German-American descent, and Mary Razor of half-Chippewa parentage. Bender grew up at White Earth Reservation in Minnesota and attended Lincoln Institution in Philadelphia, Pennsylvania, from 1892 to 1896. He played football and baseball at Carlisle Indian School in Pennsylvania from 1898 to 1901 and enrolled at Dickinson College in 1902, pitching that summer as "Charles Albert" for the Harrisburg (Pennsylvania) Athletic Club to help pay his tuition. Philadelphia Athletics scout Jesse Frisinger saw him defeat the Chicago Cubs in an exhibition game.

Nicknamed "Chief," Bender joined the Philadelphia Athletics in 1903. Over the next eleven seasons, the six-foot two-inch, 185-pound right-hander won 191 games for Philadelphia. Bender relied on a good fastball, a sharp curve, excellent control, and keen intelligence. After finishing 18–11 in 1905, he hurled a four-hit shutout in game two for Philadelphia's only World Series win over the New York Giants. Although surrendering only five hits in game five, Bender lost, 2–0, to Christy Mathewson. His best season came in 1910, when he led the American League with a 23–5 won-lost record and an .821 winning percentage, allowed only forty-seven walks in 250 innings, and posted a 1.58 ERA.

In 1911, Bender logged a 17–5 record and posted two victories in the fall classic against the New York Giants. After Mathewson edged him, 2–1, in game one, he bested Mathewson, 4–2, in game four and won decisive game six in a 13–2 rout. In 1913, Bender finished 21–10 to help Philadelphia win another American League pennant, winning fifteen of twenty-one starts and six games in relief and saving a league-high thirteen games. Bender defeated the New York Giants, 6–4, in game one and 6–5 in game four to help the Athletics win another fall classic. In 1914, he paced the National League in winning percentage with a 17–3 record and won fourteen consecutive games. He lost the opening World Series game, 7–1, to the Boston Braves, who swept Philadelphia in a major upset.

The financially strapped Athletics let Bender jump to the rival Federal League, which had claimed major league status in 1914. Bender signed with the Baltimore Terrapins for a $5,000 bonus and $8,500 salary, compiling a 4–16 mark in 1915. The next two seasons, he again pitched in Philadelphia, but this time for the Phillies. Bender then left the major leagues, except for a one-inning stint with the Chicago White Sox in 1925.

After working in a shipyard in 1918, Bender managed Richmond, Virginia, of the Virginia League in 1919 and won twenty-nine of thirty-one decisions on the mound. He pitched and managed at New Haven, Connecticut, of the Eastern League in 1920–1921; Reading, Pennsylvania of the International League in 1922; and Johnstown, Pennsylvania, of the Middle Atlantic League in 1927. Bender coached for the Chicago White Sox in 1925–1926, the U.S. Naval Academy in 1928, and the New York Giants in 1931. He managed the independent House of David during the 1930s, Erie, Pennsylvania, of the Central League in 1932; Wilmington, Delaware, of the Interstate League in 1940; Newport News, Virginia, of the Virginia League in 1941; and Savannah, Georgia, of the South Atlantic League in 1946. Bender scouted for the Philadelphia Athletics in 1945 and 1947 through 1950, and coached there from 1951 through 1953, when cancer and a heart condition forced him to retire. In 1953, he was elected to the National Baseball Hall of Fame.

Bender married Marie Clements of Detroit, Michigan on October 3, 1904. They had no children. Bender engaged in the watch making, jewelry, and clothing businesses, painted landscapes in oils, and proved an excellent marksman, trap shooter, golfer, and billiards player.

David Porter

FURTHER READING

Lieb, Fred. *Connie Mack: Grand Old Man of Baseball.* New York: G.P. Putnam's Sons, 1945.

Mack, Connie. *My 66 Years in Baseball.* Philadelphia: Winston, 1950.

Tholkes, Robert. "Chief Bender: The Early Years." *Baseball Research Journal* 12 (1983): 8–13.

Martha BENJAMIN

Born 1935
Cross-country skier

Martha Benjamin is a lifetime resident of Old Crow in the Yukon, a tiny Arctic community on the banks of the Porcupine River about 300 miles northwest of Whitehorse. It is home to the Vuntut Gwitchin

First Nation. Born in 1935, she grew up in a traditional native community sustained by hunting and trapping. She married young and had five children by the time she was twenty-five. Always athletic but with little opportunity to play sports, Martha took up cross-country skiing under the guidance of Father Jean Marie Mouchet, an Oblate priest who had served with the French ski troops during the Second World War. In the late 1950s, he recognized the potential for encouraging and producing top-level skiers in the MacKenzie Delta, and on his own set up a training program for those interested.

Because of the isolated location of Old Crow, it was difficult for Father Mouchet's skiers to compete except among themselves, although he got them to a competition in Fairbanks, Alaska, where, the *Whitehorse Star* reported on January 28, 1963, they had "no trouble cleaning the field." In that month he also managed to bring several men and women (including Martha) to Whitehorse so they could train for the Canadian Nordic Ski Championships (Western Division) in Revelstoke, British Columbia, with a view to putting several Old Crow skiers on the Olympic team. Whitehorse businesses and residents responded generously, and in the end some $3,000 was raised to aid the skiers. Martha, along with Ben Charlie, flew to Ottawa to train and compete in the U.S. nationals in Franconia, New Hampshire, the North American Ski championships in Crested Butte, Colorado, and finally the Canadian championships in Midland, Ontario.

Aside from races in Alaska, this was the first time Martha had competed against top North American skiers. In the early 1960s there was little interest among Canadian and American women in cross-country ski racing, and those who did compete came mostly from Finnish or Norwegian backgrounds. Sometimes there were no other women competing at all, as was the case at the U.S. Nationals in New Hampshire, where Martha came twenty-sixth in a field of sixty-nine. She had new boots, skis, and correct fittings, supplied through Indian Affairs. At the Canadian championships in Midland, racing against women, Martha won the ten-kilometer event with a time of 43.29 minutes. She crossed the finish line nearly a minute and a half ahead of Anni Saari and Ula Jamsa, two more experienced skiers from Sudbury, Ontario, and was declared the official 1963 Canadian women's Nordic champion.

Returning to Old Crow, Martha continued to train, hoping to make the Canadian Olympic team. Unfortunately, there was no money and little interest on the part of the Canadian Olympic Committee in sending her to the 1964 Winter Olympics in Innsbruck, Austria. Although the women's Nordic events had been part of the Olympic program since 1952, it was not until 1972 at the Sapporo Olympics in Japan that Canada (and the United States) sent a women's team. By then, the Territorial Experimental Ski Training (TEST) program had been established, assisted by funds from the federal government and based on Father Mouchet's original project of training skiers in the MacKenzie Delta.

Today, Martha is a lively, active woman with thirteen grandchildren. She has her own dog team and loves to spend time at her cabin in the bush. She still skis and helps maintain the trails around Old Crow. She is a member of the Canadian Rangers, part-time reservists who provide a military presence in remote, isolated, and coastal Canadian communities, and was recently awarded a Special Service medal by the governor general. In 1989, she was inducted into the Sport Yukon Hall of Fame for her accomplishments in cross-country skiing.

M. Ann Hall

FURTHER READING

Benjamin, Martha. Telephone interview with M. Ann Hall, May 7 and June 7, 2002.

Bruce, Harry. "Martha Benjamin: She Skis Like a Pro, When She Isn't Shooting Caribou." *Maclean's*, December 14, 1963, 66.

Sport Yukon, "Hall of Fame" (www.sportyukon.com/hallOfFame/mbenjamin.html).

Louis "Deerfoot" BENNETT

Born c. 1831
Died January 18, 1896
Runner

Often characterized as the "world's best runner" in contemporary accounts, Louis "Deerfoot" Bennett was a popular and dominant figure in professional distance racing in both the United States and Great Britain during the mid-nineteenth century. His accomplishments, however, were largely forgotten in the wake of the "amateurism" movement that governed track athletics from the late nineteenth to the late twentieth century.

Louis Bennett was the English name given to Hagasodoni, a Seneca and member of the Snipe Clan born on the Cattaraugus Reservation in western New York. The contemporary public knew him best as "Deerfoot," a name he reputedly earned for outrunning a horse early in his professional racing career.

The first hint of that career came in 1854, when the *New York Clipper* began to cover his performances in the Buffalo area. In 1856, Deerfoot began to capture the public's interest after posting an extraordinary win in a five-mile race at a fair in Fredonia, New York. Many doubted the accuracy of his recorded time—twenty-five minutes flat—but few could dispute his ability when in 1857 he defeated New England champion John Stetson in a ten-mile contest. Deerfoot returned to the Fredonia Fair in 1858, winning the ten-mile race in fifty-six minutes, nineteen seconds and in 1859 when he once again ran the five-mile distance—this time in twenty-five minutes, eighteen seconds. Sometime during this period, Deerfoot began running races in Massachusetts and the New York City area.

In 1861, Deerfoot ran his first "international" race in New York City, against three runners from England. Though he lost to all three, he ran well enough to pique the interest of George Martin, an English promoter who had arranged the race; he later signed Deerfoot to race in England.

The arrangement was typical of professional footraces during the nineteenth century, especially in Great Britain, where "pedestrianism" enjoyed its greatest popularity. English competitions, in particular, were better organized, the crowds much larger, and the monetary rewards more significant than in the United States. The larger contests in England were also conducted in a carnival atmosphere; many athletes adopted costumed personas. English spectators had little difficulty recognizing the Seneca from New York, who arrived for each race, at Martin's direction, wearing a wolf-skin blanket and headband with an eagle feather affixed to it. Martin even went so far as to employ an "interpreter" for his English-speaking runner.

Nineteenth-century Europe was enamored of the American Indian, depicted so romantically in contemporary popular fiction, so this "Indian" persona undoubtedly attracted some spectators. It was Deerfoot's talent, however, that really drew the crowds. After losing his inaugural race in England on September 9, 1861, Deerfoot posted an unbroken string of victories during his first four months, racing at least once a week.

When attendance at the races began to

drop, Martin arranged an exhibition tour of runners who traveled through Scotland, Ireland, and northern England contesting a four-mile race at each stop. Deerfoot won every race on this tour, but his victories were marred by the allegations of a fellow competitor traveling with the exhibition that the races were fixed—a practice not uncommon to the sport.

Fixed or not, the sum of these races, which by tour's end in September 1862 numbered over a hundred, took their toll on Deerfoot. When he returned to competition in the London area in October 1862, Deerfoot performed inconsistently, posting slow times and in at least one instance failing to finish. Spectators, who often bet not only on the outcome but the elapsed time, quickly lost interest. Within the month, however, he was back on top after a record-setting performance: on October 27, 1862, within the space of an hour, Deerfoot covered a distance of eleven miles, 720 yards.

Deerfoot's most enduring race, however, was still a few months away. On April 3, 1863, he contested a twelve-mile handicapped race in which some of England's best runners were given a head start. Deerfoot won, breaking four records in the process, including his own one-hour standard set in October of the previous year. Deerfoot's records for twelve miles (one hour, two minutes, 2.25 seconds), the eleven miles, and the ten miles (fifty-one minutes, twenty-six seconds), lasted well into the twentieth century. His new one-hour mark of eleven miles, 970 yards stood until 1897.

His best performances behind him, Deerfoot returned in 1863 to the United States, where he continued to run competitively, racing at county fairs and, after the Civil War, organizing his own traveling exhibition of runners. At the age of forty, Deerfoot retired from racing to his home on the Cattaraugus Reservation. He made his last public appearance in 1893 as part of an exhibition of Native Americans at the Chicago World's Fair.

Louis "Deerfoot" Bennett died three years later in 1896. In 1899, his remains were moved to a cemetery in Buffalo, New York, where they were interred next to the grave of the famous Seneca leader Red Jacket.

Caoimhín P. Ó Fearghail

FURTHER READING

Lucas, John. "Deerfoot in Britain: An Amazing American Long Distance Runner, 1861–1863." *Journal of American Culture* 6, no. 3 (1983): 13–18.

Lovesey, Peter. *The Kings of Distance: A Study of 5 Great Runners.* London: Eyre and Spottiswoode, 1968.

Arthur "Chief" BENSELL

Born Unknown
Died Unknown
Football player

A star right end for Heidelberg College in the early 1930s, Arthur "Chief" Bensell provided leadership on the field for the Student Princes of the Ohio Athletic Conference. His breakout 1932 season in his junior year garnered him first team All-Ohio Athletic Conference honors, and he captained his team as a senior.

Heading into the fall of 1932, fans of the football team of Heidelberg College, in Tiffin, Ohio, hoped the team would recover from a disastrous 1931 season. Local papers rated the team as "dangerous," with the potential for a good season. Bensell was one of many returning veterans that Coach Ted Turney could lean on, but no one predicted the junior lineman would be a starter. Indeed, he only played as a substitute in the first two games of the season, a tie and a loss. In the third game, a loss to Toledo, Bensell again came onto the field as a substitute. But this time he gained the attention of coaches and fans alike as he repeatedly put Toledo players on the ground and outplayed starter Vinton Blum.

The game against Toledo signaled the turning point for Bensell and his Heidelberg teammates. At a time when athletes played offense as well as defense, the Heidelberg junior excelled on both sides of the ball, and the team did not lose the rest of the way. Following a 13–0 pasting of Muskingum on October 22, local sportswriters praised "Chief" Bensell as a tower of strength and spoke of some of the "prettiest end playing" seen in the home stadium for several years. His efforts in that victory ensured his starting berth for the remainder of the season. In the final four games of the season, Bensell played every single minute in wins against Mt. Union, Ashland, Akron, and John Carroll. The victories guaranteed the Student Princes second place in their conference and solidified Bensell as a star on the best defense in the conference. In the games against Mt. Union and Ashland, Bensell also proved he could play offense. He caught an extra point in the 21–0 victory over Mt. Union, but his performance against Ashland garnered more attention. In an exciting 6–0 win, Bensell made two athletic catches in the last minute to set up a touchdown plunge by star halfback Eddie Zipfel. Headlines the following Monday proclaimed, "Chief Bensell Saves Day for Heidelbergers."

At the end of the season the United Press as well as the coaches and athletic directors in the twenty-one-team conference placed Bensell on the All-Conference team. Heidelberg further recognized his ability and leadership by naming him as the captain for the 1933 season. Bensell's senior year did not match the glory of his 1932 campaign, and he did not repeat on the All-Conference team. But he served capably as the captain for the Student Princes as they once again played spirited football against their opponents in the Ohio Athletic Conference.

John P. Bowes

FURTHER READING

Heidelberg College, "Heidelberg Athletics: Football Records" (www.Heidelberg.edu/athletics/football/records/nf-alltimeoac.html).
The Tiffin Daily Advertiser.
The Tiffin Daily Tribune.

SuAnne BIG CROW

Born March 15, 1974
Died February 9, 1992
Basketball player

In her short life, Big Crow (Lakota) achieved a great deal, leaving a legacy both within and beyond the athletic arena. Prior to her death she worked tirelessly on the basketball court, earning state and national honors for herself and her school. Off the court she worked just as diligently on issues of social justice, hoping to improve the lives of Lakota youth.

Although Big Crow enjoyed a number of activities and sports, including running, cross-country, cheerleading, softball, and volleyball, she was especially drawn to basketball. As a child she practiced dribbling and shooting for hours to hone her skills. Big Crow's hard work and determination paid off when she entered her freshmen year at Pine Ridge High School in the fall of 1988.

In her first year in high school Big Crow led the Pine Ridge Lady Thorpes basketball team to a fourth-place finish in the South

Promising young basketball talent, SuAnne Big Crow died tragically in a car accident. *(© SuAnne Big Crow/Boys and Girls Club)*

Dakota state tournament. This finish, as well as her strong performance during the season, provided clear indications of things to come. As a sophomore, in a game against Lemmon, South Dakota, Big Crow scored sixty-seven points in a single game, setting a state single-game scoring record. By the end of her sophomore year Big Crow averaged over thirty points per game and scored a record breaking 761 points for the season. She led her team to the state championship, scoring the winning basket to give Pine Ridge the state title. Following Big Crow's sophomore and junior years she was invited to tour various parts of the world with the Native American All-Star girls' basketball team. By the end of her senior year SuAnne Big Crow was a three time All-State selection and had received a *USA Today* All-American honorable mention.

In addition to Big Crow's athletic honors she ranked first in her class academically. She also participated in a campaign that took her to other parts of the United States combating drug and alcohol use among Native American youth.

Following SuAnne Big Crow's death from a car crash in 1992, her mother, Leatrice "Chick" Big Crow worked to create the kind of space for young folks on the Pine Ridge Reservation that SuAnne had spoken of during her life. The SuAnne Big Crow Boys and Girls Club was the first club of its kind to be established on an Indian reservation. Today the club provides educational and recreational activities for young boys and girls on the reservation.

SuAnne Big Crow was inducted into the Lakota Nation Basketball Hall of Fame in 2001. The National Education Association presents the SuAnne Big Crow Memorial Award to the K–12 student who exemplifies what Big Crow's life represented. Nominees must demonstrate their commitment to social equality by working to bring an end to bigotry and discrimination. The "Spirit of Su Award" is given annually by the South Dakota High School Activities Association to athletes within the state who demonstrate high academic, athletic, and community achievement. It is considered the highest honor given to a high school athlete in the state of South Dakota.

Rita M. Liberti

FURTHER READING

Frazier, Ian. *On the Rez*. New York: Farrar, Straus, and Giroux, 2000.

Sampson BIRD

Born August 14, 1885
Died January 24, 1952
Football player

Football and lacrosse player, and track and field athlete, he was a member of the great Carlisle Institute football teams with Jim Thorpe (1909–1912, 1914–1915) and regarded as one of the greatest ends that ever played for Carlisle. He was captain of the football team in 1911 and of the lacrosse team in 1910, and was named All-American Honorable Mention for collegiate football in 1911.

Born to Mattie Medicine Wolf Woman, a full-blooded Piegan, and John Bird, who was of English descent, Bird is notable for how others misinterpreted him and his heritage. Owing to a misspelling of his name by the Bureau of Indian Affairs, his name is often listed as "Burd." He was also often misidentified as Nez Percé, Crow, or Sioux.

Bird enrolled at the Carlisle Indian Industrial School in 1909 and took to competitive sports. A superb runner, he joined Carlisle's track and field team, as well as the lacrosse and football teams. His best years for track and lacrosse were 1909–1911. He was weight man for the track teams and captain of the lacrosse team in 1910.

He is best known for football, noted for his lead blocking for Jim Thorpe, executing his "trick play"—the end-around, a favorite play of the Carlisle Indians. Bird continued the team's success after taking over the captain's spot when Thorpe left the team in 1911 to train for the Olympics. Under Bird's leadership, the Indians were 11–1. That year, sportswriters voted him All-American Honorable Mention. Bird played three more seasons with the Indians, returning to the Blackfeet reservation after leaving the school in 1915.

He married Margaret Burgess, a Haida/Tlinget woman, whom he met while at Carlisle. They had seven children. After Carlisle, Bird and his brothers became successful cattle and sheep ranchers on the Blackfeet Indian Reservation. He was later elected to the Blackfeet Tribal Business Council and serve as a Glacier County (MT) deputy sheriff. He was elected to the American Indian Athletic Hall of Fame in Lawrence (KS) in 1985.

Edward W. Hathaway

FURTHER READING

Barlow, Earl J. "Sampson George Bird." Presentation at Bird's induction into the American Indian Athletic Hall of Fame, Lawrence, Kansas, March 30, 1985.

Steinbeck, John S. *Fabulous Redmen: The Carlisle Indians and Their Famous Football Teams.* Harrisburg, PA: J. Horace McFarland, 1951.

James BLUEJACKET

Born 1887
Died March 26, 1947, Pekin, Illinois
Baseball player

James Bluejacket, a Cherokee pitcher, had a brief career in major league baseball during the second decade of the twentieth century. He played for minor league teams in Keokuk, Iowa (1909–1911), Pekin, Illinois (1911–1912), and Bloomington, Illinois (1912–1914), before making major league debut for Brooklyn Tip-Tops of the Federal League on August 6, 1914. In 1916, after two seasons in Brooklyn, he joined the Cincinnati Reds. Bluejacket compiled a career record of thirteen wins and seventeen losses in forty-four big-league games. He continued his baseball career in the minor leagues. He played in Dallas, Milwaukee, Lincoln (Nebraska), St. Joseph, Clinton (Iowa), Columbus (Ohio), and Oklahoma City.

Bluejacket worked for Standard Oil in Wyoming in the early 1920s. At the same time, he organized and managed the Greybull, Wyoming, team of the Midwest Baseball League. In 1923, he fell from an oil tank and suffered injuries that ended his baseball career. Subsequently, he remained in the oil industry, spending fifteen years in Aruba, where he played an instrumental role in the introduction of baseball. Bluejacket Street in Aruba is named in his honor.

Bluejacket retired to Greybull, Wyoming and died at age fifty-nine on March 26, 1947, in Pekin, Illinois, while visiting his wife's family. His great-grandson, William Carl Wilkinson, played major league baseball for the Seattle Mariners (1985–1988).

Richard Thompson

FURTHER READING

Special Collection. Baseball Hall of Fame. Cooperstown, NY.

George BLUE SPRUCE

Born Unknown
Tennis player

George Blue Spruce is best known as the first Native American to become a dentist. As a teenager, though, he played championship tennis in New Mexico and continued to excel in the game as an adult.

Blue Spruce's father, who died when George was twelve years old, was a cabinetmaker from the Laguna Pueblo. His mother was from the San Juan Pueblo near Santa Fe, New Mexico, and worked as a cook at the Santa Fe Indian School. George attended an all-boys' school and spent his summers with his grandparents. It was his grandmother who gave George his Indian name, Fon-Ten-Bay-Stehn, which means Snow White Bow.

In high school, he played tennis and won the Santa Fe High School City Championship Tournament in 1949. Blue Spruce was successful in school, student leadership, and sports partly because he had a mentor and role model, Doc Renfro. Renfro was a dentist who volunteered in a youth program. He encouraged George Blue Spruce, and the youth decided to become a dentist as well.

Although counselors tried to discourage him, Blue Spruce went to Creighton University in Nebraska and studied dentistry, working part-time to supplement a scholarship he had received from the Elks Club. In 1956, he became the first Native American dentist.

Blue Spruce joined the service for two years, becoming the dentist of the first nuclear-powered submarine of the U.S. Navy, the USS *Nautilus*. Afterward, he began a career with the U.S. Public Health Service, working in the Indian Hospital of Fort Belknap, Montana. He returned to school in the 1960s to earn a master's degree in public health, and in 1968 through 1970 he traveled throughout South America as a consultant in dental health for the World Health Organization of the United Nations.

During the next twenty years, Dr. Blue Spruce held many important jobs. He was the first director of the Office of Health Manpower Opportunity and then became the director of the Phoenix Area Indian Health Service. He served as the assistant surgeon general of the U.S. Public Health Service, the only American Indian to hold that position. He retired, officially, in 1990.

He has won many awards, such as the Outstanding American Indian Achievement Award of 1974 from the American Indian Council Fire and the Alumni of the Year in 1984 from Creighton University.

With all his other activities, Dr. Blue Spruce has continued to play tennis. He and Noah Allen founded the North American Indian Tennis Association, which held its first tournament in 1976. The following year, Blue Spruce won the National Indian Tennis Championship. Ten years later, he won a gold medal in the Arizona Senior Olympics. In 1996, George Blue Spruce was inducted into the American Indian Athletic Hall of Fame.

"Don't let anybody stand in your way and discourage you from your goals," Dr. Blue Spruce said in 1997. "Indian people, Indian youngsters, you were born today because your people need you."

Vickey Kalambakal

See also: American Indian Athletic Hall of Fame; North American Indian Tennis Association.

FURTHER READING

Durrett, Deanne. *Healers*. New York: Facts On File, 1997.

Schroeter, Elaine. "George Blue Spruce." *News from Indian Country* 9, no. 12 (September 4, 1993): 13B.

BOARDING SCHOOLS

Arguably no institution did more to foster the development of modern athletics in indigenous communities than the off-reservation boarding school. Over a fifty-year period, these educational institutions increasingly used sports as a means to Americanize Native American youth. Even as athletics furthered federal policies of assimilation, they also offered Indian students important opportunities to redefine themselves in indigenous terms.

Education and Assimilation

Beginning in the last quarter of the nineteenth century, the federal government assumed full responsibility for educating American Indians. Off-reservation boarding schools became a key component of this effort. The rise of cultural evolution in popular thought, the nearly complete pacification of Native nations, and an increasing dissatisfaction with schools located in or near reservation communities facilitated the formation of boarding schools.

Richard Henry Pratt proved instrumental in this new phase of Indian education. Pratt came to appreciate the power of education administering a group of Indian prisoners of war. Guided by his penal experience and the dominant reform philosophy—encapsulated in the phrase "kill the Indian, save the man"—he successfully advocated the placement of Native Americans at the all-black Hampton Institute and later founded Carlisle Indian School in 1879.

Over the next twenty-three years, twenty-five off-reservation boarding schools were established. At root, boarding schools actively sought on the one hand to assimilate and indoctrinate and on the other hand to undermine and expunge indigenous beliefs and behaviors. Students often traveled great distances from home, often against their wills, and entered institutions in which they were forbidden to speak their native languages; were sometimes given new, foreign names; were forced to wear alien and uncomfortable clothing; were required to change their appearance, most notably by cutting their hair; and were taught the propriety of EuroAmerican actions, institutions, and values. Instruction stressed vocational over intellectual training, military discipline over self-expression. Young men learned trades frequently unsuited for their home communities, while young women learned

Organized sports and athletic teams like the girls basketball team at Tulalip Indian School, pictured here, became central to the mission of Indian boarding schools. *(© Museum of History and Industry)*

domestic arts. "Outing," a system of experiential learning in which students lived and worked in the white world, was developed to hasten acculturation. Increasingly, athletics, civic rituals, and arts supplemented the remedial education and manual labor at the core of the boarding school experience. One unintended consequence of boarding schools was stimulation of a pan-Indian identity.

In part because of inadequate funding and mismanagement, policy makers, advocacy groups, and the general public became dissatisfied with boarding schools in the early twentieth century. After 1900, the Bureau of Indian Affairs paid local school districts to take an ever increasing number of pupils. By 1930, the majority of American Indian students attended public schools.

Sports and Boarding Schools

Sports played a prominent role in off-reservation boarding schools. Administrators and educators increasingly incorporated physical activities as a part of the formal curriculum and developed organized teams that participated in interscholastic competition. In the process, those boarding schools created elite athletes recognized as all-Americans, as well as athletic powerhouses, first at Carlisle and later Haskell, that dominated intercollegiate sports. At the same time, less formal, pickup games became a feature of everyday life at the schools.

Sport became central to efforts to assimilate Native Americans. Indeed, it was understood as a means to cultivate the development of American ideals and values, including discipline, character, and citizenship. Importantly, sports not only contributed to the reconfiguration of indigenous individuals and communities but fit into broader schemes to publicize the progress of assimilation. The play of Native athletes on the football field or the basketball court allowed administrators to highlight how well students were adapting to American society and endorsing its values, including competition, sportsmanship, and teamwork. While the broader public may have accepted such progressive statements, they seemed more drawn to the spectacle of such sporting competitions. Games between Carlisle or Haskell and elite institutions like Harvard and the University of Chicago attracted huge crowds and intense media coverage. They offered occasions for the repetition of racial stereotypes, for recycling clichés about the interracial struggle for the frontier, and—not uncommonly—for remarks upon the superiority of white civilization.

For Native American students, sports were something more and altogether different than what administrators, spectators, or journalists made of them. Football, basketball, baseball, and lacrosse, as well as other sports, often provided a time out from the pressures of study and the oppression of institutional life so intent on quashing individuality and heritage. Beyond pleasure, organized sport granted Native athletes opportunities to travel and see places that might otherwise have been closed to them.

At the same time, athletics were inseparable from identity. Sports were a space in which indigenous students made statements about who they were as individuals and as Indians. Pickup games at school often became competitions between ethnic groups in which members of one tribe would endeavor to best members of another tribe, to claim honor and superiority.

In organized competitions with white schools, Native American athletes endured racism from fans and athletes but often turned stereotypes about Indians to their advantage. They used their participation in sports, moreover, to make powerful statements of ethnic pride. Finally, for many indigenous athletes, boarding schools led to successful careers as players and coaches.

C. Richard King

See also: Assimilation; Carlisle Indian School; Chewama Indian School; Haskell Indian School.

FURTHER READING

Bloom, John. "'Show What an Indian Can Do': Sports, Memory, and Ethnic Identity at Federal Indian Boarding Schools." *Journal of Indian Education* 35 (1996): 33–48.

———. *To Show What an Indian Can Do: Sports at Native American Boarding Schools.* Minneapolis: University of Minnesota Press, 2000.

Child, Brenda J. *Boarding School Seasons: American Indian Families, 1900–1940.* Lincoln: University of Nebraska Press, 1998.

Oxendine, Joseph B. *American Indian Sports Heritage.* 2d ed. Lincoln: University of Nebraska Press, 1995.

Peshkin, Alan. *Places of Memory: Whiteman's Schools and Native American Communities.* Mahwah, NJ: Lawrence Erlbaum, 1997.

Phyllis "Yogi" BOMBERRY

Born 1943, Six Nations Reserve, Ontario, Canada
Softball player

Phyllis "Yogi" Bomberry, whose nickname reflects her superb catching ability on the fast-pitch softball diamond, is from the Six Nations Reserve in southwestern Ontario, near Brantford, the largest aboriginal community in Canada. Born in 1943, she is the second oldest in a family of four brothers and two sisters, members of the Cayuga tribe, Wolf Clan. Looking back on her childhood, she readily traces the reason for her early competitiveness and prowess in sports. "To be 'in' with the kids, you had to play with the guys," which meant a steady rough-and-tumble diet of hockey, snowsnake (an indigenous sport), football, lacrosse, even hunting and shooting, and of course hardball. She recalls catching ball for her father and brother, both amateur players, who needed to practice their pitching skills. In school, she eagerly participated in basketball, badminton, gymnastics, and volleyball, and in house leagues both on and off the reservation she played softball.

The Ohsweken Mohawk Ladies Softball team, formed in 1945, was the major outlet for talented women ball players on the Six Nations Reserve. By the late 1950s Phyllis was experienced enough to join the team, and helped them win the Intermediate B Provincial Women's Softball Union championships in 1960 and 1962. She played with the Ohsweken Mohawks until 1963, when she moved to Toronto to attend high school. The Toronto Carpetland Senior A team, which played in the Ontario senior women's league, saw an obvious talent and helped Phyllis obtain employment so she could play ball for them. The team won the senior Canadian women's softball championships in 1967 and 1968. Phyllis was also top batter at the 1967 championships and the All-Star catcher for Canada in 1967 and 1968. She was on the Ontario senior women's championship team in 1967 and 1968, when it won the gold medal at the Canada Games in 1969. Phyllis personally won the Canadian All-Star catcher and Most Valuable Player awards.

She continued to play for Toronto Carpetland for just another year, when she was forced to retire from competitive ball after twenty-two years due to a serious knee injury. With pride, she points out that her operation was on the same day as that of Bobby Orr, the famous NHL hockey player, whose career ended the same way.

In recognition of her sporting accomplishments, Phyllis was the first female recipient of the Tom Longboat Award, given annually (since 1951) to the most outstanding Canadian aboriginal athlete, male or female; she won it in 1969.

Today, Phyllis is retired and lives on the Six Nations reserve in Ohsweken. Unfortunately her injured left knee has prevented her from actively playing and competing in sport for many years. Instead, she enjoys watching almost all sports on television, especially football, and pursuing a longtime interest in native artwork and crafts.

M. Ann Hall

FURTHER READING

Koserski, Lucy. "Indians Triumph." *Spectator,* June 6, 1969, 45.

Paraschak, Vicky. "An Examination of Sport for Aboriginal Females of the Six Nations Reserve, Ontario from 1968 to 1980." In *Women of the First Nations,* edited by Christine Miller and Patricia Chuchryk. Winnipeg: University of Manitoba, 1996.

Henry Charles BOUCHA

Born June 1, 1951, Minnesota
Hockey player

Henry Boucha was an outstanding hockey player whose promising career was cut short by injury. He was born outside of Warroad, Minnesota, just north of the Canadian border in 1951. His mother was Ojibway (Chippewa), and his father was Ojibway, Cree, and French. Boucha grew up in Warroad playing football and baseball, as well as hockey. His sixth-grade bantam hockey team won a state championship.

In high school, Boucha was voted to the Minnesota All-State team three years—the first person in state history to receive that honor as a sophomore. During his senior year he led tiny Warroad to the state tournament in St. Paul. The final of the tournament featured Warroad against a team

A gifted defenseman, Henry Boucha saw his hockey career cut short by injury. *(© Bruce Bennett Studios)*

from Edina, a suburb of Minneapolis. Edina was a large school with wealthy, urban students, and the game became legendary in the annals of Minnesota sports. Edina was heavily favored but managed to win by only a score of 5–4 in overtime. The tenacity of the Warroad team and its star player, however, made Henry well known throughout the state.

Boucha received several offers to play hockey after high school. He turned down a scholarship at the University of Minnesota to play junior hockey in Winnipeg. A defenseman throughout high school, Henry was switched to center and played that position for the remainder of his career. In 1970 he accepted an invitation to play on the U.S. Hockey National Team. This group competed in international competition for two years, culminating in a trip to the 1972 Winter Olympics games in Sapporo, Japan. The sixth-seeded U.S. team surprised everyone by winning a silver medal. Henry Boucha scored two goals and four assists in the six-game Olympics tournament.

After the Olympics, Boucha signed with the Detroit Red Wings of the National Hockey League, scoring a goal in his first NHL game. The flamboyant Boucha grew his hair long and kept it out of his eyes by wearing a white headband instead of a helmet. He proudly embraced his nickname of "Chief," although some Native American groups complained that it was a demeaning sobriquet. In 1974 he was traded to the Minnesota North Stars.

On January 4, 1975, Boston Bruins player Dave Forbes got into a fight with Boucha and smashed the butt end of his stick into Henry's eye. The blow broke his eye socket and required thirty stitches to close. The game had been played in Minnesota, and the local district attorney filed assault charges against Forbes—the first time such action was ever taken by civil authorities for activities that occurred in a sporting contest. Forbes's trial ended in a hung jury, and the prosecutor decided not to retry him. Boucha sued Forbes, the Bruins, and the NHL, receiving a reported $1.5 million settlement.

Boucha played two more years (including part of one in the upstart World Hockey Association), but he suffered double vision, which limited his effectiveness. He retired shortly after the 1976–1977 season began—at the age of twenty-six. In 1995 he became the first person of Native American descent to be inducted into the U.S. Hockey Hall of Fame.

Roger D. Hardaway

FURTHER READING

Gilbert, John. "Warroad's Boucha Wins Election to U.S. Hockey Hall of Fame." *Minneapolis Star-Tribune*, August 12, 1995.

Schofield, Mary Halverson. *Henry Boucha: Star of the North*. Edina, MN: Snowshoe, 1999.

Emmet Jerome BOWLES

Born August 2, 1898, Wanette, Oklahoma
Died September 3, 1959, Flagstaff, Arizona
Baseball player

Bowles attended school in Wanette, graduating from its high school. From 1917 to 1919 he served in the U.S. Army, fighting in World War I. He also attended Sacred Heart College (now St. Gregory's University) in Shawnee, Oklahoma, for one year.

He debuted as a Chicago White Sox pitcher on September 12, 1922, when he was twenty-four years old. Bowles, a right-handed pitcher, was six feet tall and weighed 180 pounds. His nickname was "Chief." In his debut (and his only recorded major league game), he pitched one inning, giving up two hits and one walk. Three runs scored. Bowles then pitched for semipro clubs, including Western and Southwestern League teams, while continuing to look for major league opportunities.

Bowles married Nona Mary Kirkham on April 27, 1927. He continued to play semipro baseball through 1940, and he also worked as a miner foreman and as a carpenter.

On September 3, 1959, Bowles was preparing to speak before the Flagstaff, Arizona, chapter of Alcoholics Anonymous. He walked toward the podium and collapsed, dying of a heart attack before deputies arrived at the site. At the time of death, Bowles was sixty-one and had been living in Albuquerque, New Mexico. He is buried in Mount Calvary Cemetery, survived by his widow, three daughters, seven grandchildren, two sisters, and a brother.

Kelly Boyer Sargent

FURTHER READING

Burke, Bob, Kenny A. Franks, and Royse Parr. *Glory Days of Summer: The History of Baseball in Oklahoma.* OK: Oklahoma Heritage Association, 1999.

BOX LACROSSE

Unlike field lacrosse, the sport of box lacrosse did not originate among Native Americans. Instead, it was the product of Canadian ice hockey promoters who wanted to fill their empty arenas in summer with paying spectators. Joe Cattaranich and Leo Dandurand, owners of the Montreal Canadiens pro hockey team, helped to create the new game by taking field lacrosse indoors and reducing the number of players from twelve to seven. They recruited field lacrosse veterans as well as ice hockey players seeking to stay in shape in the off-season. Some Canadians called the new game "summer hockey." Because these early box lacrosse contests took place in converted ice hockey arenas, the most obvious unique features of the new sport were the cement playing surfaces and the walls around the rinks.

Official play began in 1931, when promoters sponsored an International Lacrosse League. The league's four franchises included the Montreal Canadiens, Montreal Maroons, Cornwall Colts, and Toronto Maple Leafs. Insufficient crowds led to the collapse of the league during its second season. An even shorter-lived American Box Lacrosse League played in 1932 in Toronto and cities in the United States. These two pro circuits may have failed to catch on among sports fans during the Great Depression, but amateur field lacrosse clubs across Canada opted to switch over to the new box lacrosse format. If clubs were unable to find an empty hockey arena to play in, they constructed enclosed fields in outdoor playing facilities. Sometimes referred to as "boxla," the game maintained loyal followings in southwestern British Columbia and southern Ontario, though it never quite developed the level of popularity enjoyed by field lacrosse prior to the First World War.

The conversion from field lacrosse to "boxla" later, however, was fairly rapid, as the Canadian Lacrosse Association, the Ontario Lacrosse Association, and the British Columbia Lacrosse Association all adopted the new game. Beginning in 1953, the box lacrosse squad shrank from seven men to six. From the 1930s through to the present, the OLA has sponsored competition at a variety of levels including leagues for adults to those for youth. In British Columbia, the Senior clubs of the Inter-City Lacrosse League included teams representing Vancouver, New Westminster, Victoria, and Nanaimo. During the late 1960s and early 1970s, these clubs turned professional.

However, after the collapse of those leagues, the clubs reverted to amateur status and reorganized as the Western Lacrosse Association for the 1973 season. There are currently four major box lacrosse championships held in Canada: Mann Cup (for Senior A "elite" teams), President's Cup (Senior B "open"), Minto Cup (Junior A), and Founders Trophy (Junior B).

From the very beginning, Native athletes embraced this new version of lacrosse, which permitted more bodily contact than the field game. Not only did many reservation communities create and operate their own clubs, but many individual Indian athletes have played on non-Native clubs as well. By the 1990s, there were two Senior B level leagues composed of Native clubs: the Can-Am League (in Ontario and western New York) and the Iroquois Lacrosse Association (in the Saint Lawrence River Valley and central New York). Several Native clubs have won major championships. The Six Nations Chiefs won the Mann Cup in three straight seasons from 1994 to 1996. For the President's Cup, champions included the Oshweken Warriors in 1964, the North Shore Indians in 1993 and 2001, the Tuscarora Thunderhawks in 1994, the Akwesasne Thunder in 1995 and 1997, the Oshweken Wolves in 1996, and the Newtown Golden Eagles in 2000. As for Junior titles, winners included the Six Nations Arrows, who captured the Minto Cup in 1992, and the Six Nations Red Rebels, who won the Founders Trophy in 1997.

Donald M. Fisher

See also: Lacrosse.

FURTHER READING

Fisher, Donald M. *Lacrosse: A History of the Game.* Baltimore: Johns Hopkins University Press, 2002.
Hinkson, Jim. *Box Lacrosse: The Fastest Game on Two Feet.* Radnor, PA: Chilton, 1975.
Tewaarathon [Lacrosse]. Akwesasne, ONT: North American Travelling College, 1978.

BOXING

Within mainstream America boxing has been held up since before the Civil War as a manly and magnificent spectacle. Native Americans have proven themselves within amateur and professional boxing. Initially, boarding schools played a fundamental role in the promotion of boxing among American Indians. Later, amateur organizations, particularly the Golden Gloves and the Amateur Athletic Union, granted opportunities for indigenous boxers, some of whom left a mark in the professional arena. Most recently, the Native American Sports Council has sought to develop boxing talent in Indian country.

Boarding Schools

Although boxing was not central to Native American boarding schools at their inception, over time instructors and administrators at select institutions seized upon boxing as an important medium to develop character, discipline, and physique. The fiscal hardships of the Great Depression seem to hasten the rise of boxing. Football demanded fiscal support and strong player support. Boxing lent itself to reduced funding and fewer numbers of athletic personnel; by the onset of World War II (1939), boxing had assumed a leading position. It became "a core aspect of life at many boarding schools," and students saw in the sport rare opportunities of enhancing "pride and pan-Indian nationalism."

At Chilocco Indian School in 1932, a sports promoter from Wichita, Kansas, persuaded the superintendent to send a team to take part in an American Legion tournament. A year later the school boxing team took part in a regional and national competition and was profiled by no less than *Ring* magazine, the world's most highly regarded boxing magazine. Boxing success was not restricted to Chilocco; schools from Albuquerque, Haskell, Phoenix, and Santa Fe sent their champions to take part in AAU competitions.

Boxing caught on in the 1930s not just because of institutional changes but because it was an activity "that resonated with the lives of boarding school students." For students one of the attractions of boxing was that it proved an exit from the confines of school life and an access, however transitory, to the outside world. National championships after all took place in such cosmopolitan centers as Chicago, New York, and Boston. Student boxers recalled the pleasure of being able to mix and mingle socially, in contrast to a school life where the routine and ritual was one of isolation.

Boxing also provided an arena where racial pride could be appropriately displayed. As a Navajo representing a Santa Fe Indian School in the 1930s remarked, "Because you're an Indian, you're going to show what an Indian can do."

In 1948 the BIA banned boxing, but the decade of the 1930s had helped the sport craft and shape fresh ethnic personas for Native American boarding school students. "The huge popularity of the sport among students, as well as its strong association with pride rooted in the common historic experiences of Native American people, suggest that for some, boxing made it possible to understand expressions of cultural memory in diverse ways."

Noteworthy Boxers

A number of Native Americans have left their mark on modern boxing. Several have won amateur titles. In 1945, Amos Aitson (Kiowa) won the fifth annual national boxing championships (118-pound class) at Boston Garden. The same year, Virgil R. Franklin (Arapaho-Kiowa) who held a succession of boxing titles and took the 1945 National Golden Gloves championship (126 pound class). Nelson B. Levering (Omaha-Bannock) was the 1947 Midwest Golden Gloves champion.

Others have put together impressive careers. Gordon A. House (Navajo-Oneida) had a long and successful amateur career; he is perhaps best remembered for fighting featherweight champion Sandy Saddler in 1949. During a shorter career, Alvin LeRoy Williams (Caddo) established a career record of fifty-four wins (twenty-four by knockout), seven draws, and thirty-two losses. J. Earle Keel (Chikasaw) fought as an amateur and professional boxer. His record in the latter category was sixty-seven fights and only fourteen losses. In his later life he coached and refereed fights that featured Sonny Liston, Cleveland Williams, George Foreman, and Cassius Clay. Most recently, Danny "Little Red" Lopez was arguably the most successful Indian boxer of the twentieth century, reigning as world featherweight champion from 1976 to 1980.

Scott A.G.M. Crawford

FURTHER READING

Bloom, John. *To Show What an Indian Can Do: Sports at Native American Boarding Schools.* Minneapolis: University of Minnesota Press, 2000.

Oxendine, Joseph B. *American Indian Sports Heritage.* Champaign, IL: Human Kinetics, 1988.

Ellison Myers "Tarzan" BROWN

Born 1914
Died 1975
Runner

Brown grew up in a large, very poor family. He and his six siblings lived in a small,

Ellison Brown is best remembered for his victory in the fortieth Boston Marathon. *(AP/Wide World Photos)*

ramshackle hut outside of Westerly, Rhode Island. His Narragansett tribal name was Deerfoot, for his running prowess, but his nickname was "Tarzan" because of his impressive build and because he enjoyed swinging from tree branches as a boy.

Tarzan Brown ran his first important race in 1933 at the Boston Marathon, finishing thirty-second. After repeating that finish in 1934, he expected better results in 1935. But the race was only two days after his mother's death; he ran in a shirt made from one of her dresses. His shoes fell apart at twenty-one miles, and he finished barefoot in thirteenth place. But by 1936 he was well known and was sponsored by the Tercentenary Committee of Providence, Rhode Island. Able now to train without working or financial worries, Brown won the 1936 Boston Marathon by almost two minutes, leading at all checkpoints. The victory qualified him for the 1936 U.S. Olympic team. At Berlin, Brown suffered a hernia, but he started the marathon and ran with the leaders for thirty kilometers, although he failed to finish.

Later in 1936, Tarzan Brown won the venerable Port Chester marathon in 2:36:56.7. On the next day, October 12, 1936, he won a second marathon in as many days, winning the New England Marathon Championship in Manchester, New Hampshire. Brown did not fare as well in 1937, racing poorly at Boston and finishing thirty-first. The next year was even worse, as he finished the 1938 Boston Marathon in fifty-first place, running only 3:38:59. But he did finish the year by winning the AAU twenty-five-kilometer title. At the AAU Championships in Yonkers in November, he led for fifteen miles, only to fail to finish, due to abdominal cramps.

In 1939, Brown fully regained his previous form, winning Boston in 2:28:51 for a new course and American record. But in 1940, he did not finish the most important American marathon, which was won by the Canadian, Gerard Côté. Côté broke

Brown's course record, but in the next month, at the Lawrence (MA) to Salisbury Beach (NH) marathon, Brown ran his career-best time, defeating Côté and the top four finishers from the 1940 Boston, finishing in 2:27:29.6, a new American record for the marathon. He did not retire, finishing Boston in 1943 and 1946, but he was never again of international caliber.

Tarzan Brown made his living as a stonemason, and settled near his boyhood home in Charleston, Rhode Island. He died in 1975 when he was struck by a van while walking along a street.

Bill Mallon

Louis R. BRUCE, SR.

Born January 16, 1877, St. Regis, New York
Died February 9, 1968, Ilion, New York
Baseball player

Louis Bruce (Mohawk) was a professional baseball player in the very early twentieth century. He later became a Methodist minister and is also remembered as the father of Bureau of Indian Affairs commissioner Louis R. Bruce, Jr., who served under President Richard Nixon.

Louis, or Lou, Bruce, was born in St. Regis, New York in 1877. His father, John, is said to have been a chief who went to the Sudan in 1884 to help the British troops there with his boating expertise. According to a biography about his son, Lou Bruce studied dentistry at the University of Pennsylvania and paid his tuition by playing professional baseball.

Bruce first played with a minor league team, the Toronto Maple Leafs, in 1902—the year they won the pennant in a tight race with the team from Buffalo, New York. Bruce, at five feet, five inches tall, was one of three starting pitchers for the Maple Leafs and won eighteen of the twenty games he pitched. He also played the outfield in fifty-five games. His batting average for that year was .313, and he had ninety-nine hits, two of them home runs.

In 1904, Bruce moved up to the major leagues and played for the Philadelphia Athletics under the leadership of Connie Mack. One of his teammates was pitcher Charles Bender (Chippewa), later inducted into the Baseball Hall of Fame. Bruce played his first game for the Athletics on June 22, 1904, when he was twenty-seven years old. He played in only thirty games that year, pitching in two of them. In the rest of the games, he played outfield—left, center, and right—even second and third base. His team came in fourth.

Bruce placed twentieth on the list of top batting averages that year, but because he played so few games, it is difficult to say how good a player he might have become. Bruce's lifetime ERA was 4.91, and his lifetime batting average was .267. His baseball career was brief, and he had other ambitions.

Bruce married Nellie L. Rooks (Oglala Sioux). Their son, Louis R. Bruce, Jr., was born in 1906 at the Onondaga Reservation in New York and served as BIA commissioner in the Nixon administration.

Bruce apparently practiced dentistry for awhile, although he did not hold a degree, and later he became a Methodist minister. He served as pastor at both the Onondaga Reservation and St. Regis. He is remembered as a stern man who encouraged individual achievement and was strongly against drinking alcohol. Louis Bruce, Sr., died in 1968, in Ilion, New York.

Vickey Kalambakal

Pete BRUISED HEAD

Born c. 1897
Died June 1972
Rodeo rider

A Kainai, Bruised Head was born to Iiyiksina and Ootahkoi'ksissakii Takes the Gun Strong. He was a notable Canadian rodeo champion. He had several names throughout his childhood. After his parents separated, his mother, pregnant with him at the time, married Piyotsokaipii. Raised by his grandparents Many Bears and Stahtsisttayaaki, he took their name.

He was named Peter John Many Bears by the priests at the Catholic Mission Residential School, but his identity changed again soon after arriving at the Dunbow Industrial School, where the priests called him Peter John Hunting. Upon his stepfather's death, his mother married Iinokihkinii, which was mistranslated as Bruised Head. The curriculum and agenda at Dunbow Industrial School was effective, for when Bruised Head returned home, he denied his traditional Native religion.

Older than many of his fellow competitors, Bruised Head competed in rodeo events for thirteen years. One year before Bruised Head was born the first rodeo had taken place at the Blood Reserve community. Located about 120 miles south of Calgary between Fort McLeod, Lethbridge and Cardston in the Treaty Seven region, the Blood Agency is the largest reserve in Canada.

Organized by the government as part of its efforts to assimilate the Kainai, the rodeo replaced the annual Sundance. Launching a successful ranching industry in the 1890s, the Kanai, one of three tribes constituting the Blackfoot Nation, developed skills important to the sport before the first Canadian rodeo occurred at Raymond, Alberta, in 1903.

Although the Calgary Stampede was not widely attended until September 2, 1912, and not considered the official Canadian rodeo, the prowess that Indian cowboys exhibited in competitive rodeo events eased Bruised Head's way into the sport.

Bruised Head was particularly impressive at the Calgary Stampede as the Canadian roping champion in 1925 and 1927, barely missing the title a third time in 1928. Bruised Head divided his talents among rodeo events and ranked high in bronc riding both with and without a saddle, although bareback riding did not gain legitimacy within the sport until the 1950s. His trophies were permanently displayed at the Roman Catholic mission on the Blood Reserve.

He married Ootsskapinaakii Genevieve Blood. Together they had seven children. Bruised Head passed along to his descendants his skill for handling horses and other livestock, as well as his rodeo athleticism.

Rebecca Tolley-Stokes

FURTHER READING

Marsh, James A., ed. *Canadian Encyclopedia.* Edmonton, ALTA: Hurtig, 1985.
Moore, Tom. "Man About Town." *Albertan,* November 15, 1966.
Zaharia, Flora, and Leo Fox. *Kitomahkiapttminnooniksi: Stories from Our Elders.* Edmonton, ALTA: Donahue House, 1995.

Elmer BUSCH

Born 1890
Died Unknown
Football player

Elmer Busch played center and offensive guard for the Carlisle Industrial School football team between 1910 and 1914. Busch was one of two Pomo from northern

California to attend Carlisle. During World War I, he became a spokesman for his community and demanded improved conditions for Pomo living in Potter Valley, California. In the 1920s, Busch played professional football with the Oorang Indians.

Born in 1890, Busch grew up in Potter Valley, California. He entered Carlisle Indian Industrial School in 1910 and played on the football team. In 1913, Carlisle coach Glenn "Pop" Warner inserted Busch into the starting lineup at center. That year Busch was selected to the second-team All-American team. Between 1911 and 1913, Busch played on the most successful teams in Carlisle's storied history. The Indians posted a 33–3–2 record and outscored their opponents 1,097 to 226. In 1914, Busch's teammates elected him captain. Before the season concluded, though, school officials dismissed Busch from Carlisle for disciplinary reasons. During Busch's truncated 1914 campaign, Carlisle posted a disappointing 4–7–1 mark. Busch had other interests at Carlisle besides football. In 1910, he contributed an article on California Indians' use of acorns to *The Red Man*, Carlisle's student publication.

Busch was rootless after he left Carlisle. In 1916, Busch worked in Detroit in an automobile factory. In 1917 he appeared before a Senate Committee that investigated abuses at Carlisle. Busch provided testimony confirming Coach Warner's use of crude language in front of football players. During World War I, Busch returned to Potter Valley and became a community spokesman. In 1917, Busch complained to government officials that California Indians should not have to serve in World War I because of the murky citizenship laws in California and the Indians' familial duties. He told a reservation agent that the federal government "railroad[ed] Indian boys into service." The government treated Indians like citizens, he said, but Indians "could not send our children to public schools, and we couldn't put our helpless peoples in county hospital." He also protested the ineffectiveness of local teachers. Later, Busch complained to government officials about the lack of employment opportunities for Indians in Mendocino County.

In 1922, Busch played professional football for Jim Thorpe's Oorang Indians. Busch played center and guard for the team and had one interception. In 1973, the American Indian Athletic Hall of Fame elected Busch as one of its members.

William J. Bauer, Jr.

FURTHER READING

Machamer, Gene. *The Illustrated Native American Profiles.* Mechanicsburg, PA: Carlisle, 1996.

Newcombe, Jack. *The Best of the Athletic Boys: The White Man's Impact on Jim Thorpe.* New York: Doubleday, 1977.

Oxendine, Joseph. *Native American Sports Heritage.* Champaign: Human Kinetics, 1988.

Peter CALAC

Born May 13, 1892
Died January 13, 1968, Canton, Ohio
Football player

Pete Calac, a Mission Indian, attended grammar school in Fallbrook, CA, and at age fifteen was selected to attend the Carlisle Indian School in Carlisle, Pennsylvania. There, he played on the football team from 1912 to 1916, serving as team captain in 1915. Calac capably filled the void left in the backfield after Jim Thorpe departed Carlisle in 1912, and by 1915 he was regarded one of the top runners in the country. During his stint at Carlisle, he starred at fullback, defensive end, and kicker, scoring all of his team's points in several games.

Calac returned to California in 1916 and enlisted in the army. He served in the Ninety-first Division in France and Belgium during World War I, returning unharmed nearly two years later. Calac then attended West Virginia Wesleyan College, starring on the football team. While still in college, he occasionally played fullback for the Canton Bulldogs, often using the assumed name "Anderson" so his college eligibility would not be jeopardized. Along with Thorpe and several other alums of the Carlisle School, Calac helped to establish the credibility of the new professional football league.

He began a well-traveled professional football career in earnest in 1920 with the Bulldogs. It was the busiest season of his career, as he led the team with 243 yards rushing and scored three touchdowns. He played for the Cleveland Indians for most of the following season, before being signed by the Washington Senators late in the year. Senators coach Jack Hegerty signed Calac and several other former Bulldogs in hopes of defeating Canton in their annual game. Calac did not play in the game, and Hegerty's plan failed.

In 1922, Calac was reunited with his old Carlisle teammate Jim Thorpe, with the Oorang Indians. Though Calac played well at defensive end and fullback, missing only one game in two seasons, the team won only four games. In 1922, Calac enjoyed perhaps his finest professional season, averaging over eleven yards per rush and nineteen yards per catch. He also scored two of the team's six total touchdowns.

When the Oorang Indians disbanded after the 1923 season, Calac joined the Buffalo All-Americans in 1924. His impact as a player declined, as he carried the ball only nineteen times that season, though still scoring two touchdowns. He played two more seasons, returning to the Canton Bulldogs, the last of which was spent as a reserve fullback, receiver, and punter. He retired after the 1926 season.

Calac, though not inducted, has been nominated for induction into the Football Hall of Fame, and in 1964 he introduced his long-time friend and teammate Jim Thorpe upon his induction.

Calac enjoyed a long and happy life after retiring from football, serving twenty-five years on the Canton police force. He also fathered three children and had seven grandchildren with his wife of thirty-six years. He died on January 13, 1968 in Canton, Ohio.

Kevin Witherspoon

FURTHER READING

Adams, David Wallace. "More than a Game: The Carlisle Indians Take to the Gridiron, 1893–1917." *Western Historical Quarterly* 32, no. 1 (Spring 2001): 25–53.

Steckbeck, John S. *The Carlisle Indians and Their Famous Football Teams.* Harrisburg, PA: J. Horace McFarland, 1951.

Rick CAMP

Born June 10, 1953, Trion, Georgia
Baseball player

Rick Camp was a major league baseball pitcher who spent nine seasons with the Atlanta Braves. Born in 1953, in Trion, Georgia, Rick Camp attended West Georgia College. At twenty-three years of age, the six-foot one-inch, 185-pound right-hander made the major leagues on September 15, 1976. His money pitch was a sinker. Working out of the Atlanta bullpen, Camp would become a workhorse for the team, appearing in 414 games over the next few years and pitching 942 innings, with a record of 56–49 and an ERA of 3.37. Even for a pitcher, however, Camp was a poor batter, having a career average of .074.

After two full seasons in the big leagues, mostly in middle relief, Camp had to sit out the 1979 season with arm problems. He returned the next season as the closer. To the surprise of most observers, he made a personal record seventy-seven appearances and saved twenty-two games, with an ERA of 1.91. In 1981, he continued to close games, saving seventeen, and tied for the National league lead in relief wins with nine. The next season his career began to slide. While the Braves were division champions, Camp was used in a variety of pitching roles. For the first time in his career, the Braves tried to use him as a regular starter. Previously he had only started five games, but he started twenty-one games in 1982. While he was able to win eleven games, he had thirteen losses. He was only able to grab five saves that season, and just three more over his last three seasons. His ERA increased from 1.78 to 3.65. In his one postseason the Braves, with little success, continued to use Camp in the starting rotation. He would start thirty-seven games the next two seasons and made thirty-four relief appearances. The strangest moment of his later career happen on July 4, 1985, in a rain-delayed game against the New York Mets—a game that lasted six hours and ten minutes, nineteen innings, before the Mets were able to win it 16–13. With the Braves down by a run to the Mets, 11–10, Camp was sent to the plate, because the team had no more position players to pinch-hit for him. With two strikes on him from Tom Gorman, Camp hit the only home run of his career at 3 A.M. No pitcher in major league history had ever hit a home run that late in a game. While his 1985 season was not poor by major league standards, the Braves decided to purge their pitching staff on April 1, 1986, and released Camp, along with Pascual Perez, Len Barker, and Terry Forster.

T. Jason Soderstrum

FURTHER READING

Cunningham, George. "Rick Camp: The 'Big Stopper' for the Atlanta Braves." *Baseball Digest* 40 (December 1981): 58–60.

Fraley, Gerry. "No Joke: Braves Stage April Fools' Massacre." *Sporting News,* April 14, 1986, 15.

Minshew, Wayne. "Camp's Something More than Native of Georgia." *Sporting News,* May 14, 1977, 11.

Picking, Ken. "Braves Can't Break Camp, and Neither Can Opposition." *Sporting News,* August 30, 1980, 28.

Benjamin Nighthorse CAMPBELL

Born April 13, 1933, Auburn, California
Judo champion, politician

Campbell, whose paternal grandmother was Southern Cheyenne, is best known as a U.S. Senator from Colorado, the first Na-

Benjamin Nighthorse Campbell, United States Senator and former Olympian, addresses the 1999 Olympic Day Ceremony. *(AP/Wide World Photos)*

tive American to hold such a position in over sixty years. Long before becoming a politician, he was a judo champion, appearing in over 1,500 matches. In fact, Campbell claims that he owes the sport his life. As a teenager, he was in gangs and going in the wrong direction, but the sport kept him out of trouble.

After serving in Korea in the U.S. Air Force as a military policeman, Campbell received his bachelor of arts from San Jose State University in 1957. He learned judo in Korea, Japan, and throughout his university career. From 1956 until 1964 Campbell was a Pacific AAU champion in judo. In 1963 he was a gold medalist at the Pan-American games. He was the U.S. champion 1961–1963. As captain of the American team in 1963, he toured England, France, Belgium, Mexico, Holland, Japan, Taiwan, and Brazil. Campbell captained the first U.S. Olympic team at the 1964 Olympics and was the highest ranking member of the team, finishing fourth.

Teaching judo after his retirement from active competition, his Sacramento, California, judo club won thirteen team championships. He was president of the Northern California Judo Association in 1966 and 1967 and was named to the executive board of the U.S. Judo Federation in 1968. He has been a director and the secretary of the U.S. National Judo Championships.

After teaching martial arts, Campbell married, worked as a shop teacher, and moonlighted in law enforcement. He started making jewelry with Native American motifs. In 1970, he took the name "Nighthorse." Seven years later, Campbell and his wife moved to a ranch near Durango, Colorado, where they raised quarter horses and opened a gallery to display and sell his jewelry. He has won more than two hundred first-place and best-of-show awards for jewelry design.

Campbell was elected to the Colorado state legislature in 1983, serving until 1986. He was elected to the U.S. House of Representatives in 1987 from Colorado's Third District. As a member of the House in 1991, he won the fight to change the name of the Custer Battlefield Monument in Montana to the Little Bighorn Battlefield National Monument, legislation that honors Native Americans who died in the battle.

Elected to the U.S. Senate in 1992, Campbell is a leader in the public lands and natural resources policy. He is recognized for the passage of landmark legislation to settle Indian land rights and is in the forefront of sponsoring and fighting for legislation to protect Colorado wilderness and water rights.

Royse Parr

FURTHER READING

"Ben Campbell, 1969, Judo Player of the Year." *Black Belt Magazine* (w3.blackbeltmag.com/halloffame/html/16.html), 1997.

Biography of U. S. Senator Ben Nighthorse Campbell (www.powersource.com/campbell/default.html).

Viola, Herman J. *Ben Nighthorse Campbell: An American Warrior.* New York: Orion Books, 1993.

CARLISLE INDIAN INDUSTRIAL SCHOOL

In 1879, Lt. Richard Henry Pratt founded the Carlisle Indian Industrial School at the Carlisle Barracks in Carlisle, Pennsylvania. While the school was a failed experiment at implementing the U.S. government's policy of assimilating Native Americans into European-American culture, its athletic success was the epitome of sport glory and legend. In the midst of that bucolic south-central valley in Pennsylvania the athletic prowess of a group of superb Native American athletes impacted the American sports scene during the first two decades of the twentieth century.

A Football Powerhouse

Most significantly, the Carlisle Indians became one of the legendary teams of intercollegiate football. The legend developed not only by the feats of its great athletes but also from the football innovations introduced by a hallowed coach, Glenn S. "Pop" Warner, whom the Carlisle school's founder hired away from national football power Cornell. Warner, who coached at Carlisle from 1899 to 1904 and again from 1907 to 1914, produced athletic greatness.

While the school's football history started inauspiciously—only minimal victories from 1894 to 1898 over regional high schools, YMCAs, athletic clubs, and small colleges—over a fifteen-year period Warner's teams compiled a record of 108 wins, forty-one losses, and eight ties, defeating many of the major university football teams of the day. Their victories included wins over leading football powers of the era including Harvard, Pennsylvania, Cornell, Pittsburgh, Chicago, Minnesota, Nebraska, California, and an Army team captained by Dwight D. Eisenhower. The Carlisle school also played against such powers as Princeton and Yale.

The Carlisle teams compiled these incredible victories despite playing with only two or three substitutes and giving away a tremendous weight advantage to their opponents in almost every game they played;

Carlisle became a national football powerhouse in the first decade of the twentieth century, featuring players like Jim Thorpe, Gus Welch, and Joe Guyon. *(Library of Congress)*

as a team they usually averaged about 170 pounds. The fury of their attack, however, disrupted heavier lines, and the trauma of their tackling intimidated their opponents' backfields.

Not only did their victories trumpet the Carlisle Indians' fame, but their exploits were chronicled by the major sport writers of the early twentieth century. Later some accounts, especially colorful but not always accurate, were written by Bill Stern. His stories, many of which found their way into the ubiquitous Bill Stern sports books, spread the stories of the Carlisle Indians across several subsequent American generations.

In 1902, under Warner's tutelage, the small Native American school produced one of the all-time great teams in college football history. The backfield was composed of Jim Thorpe (Fox and Sac), Alex Arcasa (Colville), Gus Welch (Chippewa), and Possum Powell (Cherokee). The line included Pete Calac (Mission Indian) Joe Guyon (Chippewa), Roy Large (Shoshone), William Garlow (Tuscarora), Joe Bergie (Chippewa), and Elmer Busch (Pomo).

Many Carlisle players went on to play professional football. Among these were Joe Guyon, Albert Exendine, Pete Calac, Myles McLean, and the legendary all-around athlete Jim Thorpe. Exendine is enshrined in the College Football Hall of Fame, while Guyon, and Thorpe were elected members of both the College Football Hall of Fame and the Professional Football Hall of Fame.

Equally noteworthy is the impact that the Carlisle Indians and their coach Pop Warner had on the style of American football. Until Warner began to coach at Carlisle, football had been a game mainly of pushing and pulling, while running out of the straight-T formation. Players were large and beefy. Warner's Carlisle Indian teams showed that football could be a game of skill and strategy, speed and deception. They added to American football the three-point stance, pulling linemen, the single-wing formation, the double-wing formation, and the body block, quickly dubbed the "Indian block."

After the forward pass was legalized in 1906, Frank Mount Pleasant (Tuscarora) gained national attention throwing spirals to Bill Gardner (Chippewa) and Albert Exendine (Delaware). The forward pass joined the punt and the drop kick as effective weapons for not only the Carlisle Indians but for football in general.

As Warner described his football players, "This is a new kind of team." His players, he said, were "light but they're fast and tricky. Once they get into an open field, they're like acrobats, they're so hard to knock off their feet. . . . [T]hey are born lovers of the game. They have speed and skill in use of hands and feet. They also have highly developed powers of observation, handed down through generations." In a prideful boast he proclaimed, "There wasn't an Indian of the lot who didn't love to win and hate to lose, but to a man they were modest in victory and resolute in defeat. They never gloated, they never whined, and no matter how bitter the contest, they played cheerfully, squarely and cleanly."

In addition to the football team, the Carlisle Indian school fostered the beginnings of the careers of nationally famous athletes like Thorpe. Thorpe was one of the most highly decorated American Olympians and also played professional baseball and football. He was arguably the most recognized sports hero of his generation. He, along with other Carlisle Indian School greats, such as Chief Bender, the legendary baseball pitcher, led a constellation of Native Americans who blazed across the sport scene in the early twentieth century. Perhaps most amazing was the 1912 Olympic performance of two Carlisle athletes; at that year's games in Stockholm, Sweden, Thorpe and Louis Tewanima (Hopi), a long-distance runner, accumulated more points than athletes of any United States college or university.

Sport and Assimilation

While sports and sport heroes were the glories of the Carlisle Indian Industrial School, the government policy of the assimilation of the Native Americans into the American society of the late nineteenth and early twentieth centuries was a major failure. The school began as an experiment based on Lt. Richard Pratt's belief that Native Americans were capable of shedding their "savagery" and becoming productive citizens after receiving opportunities equal to those of the European Americans. It led to a national policy move away from annihilation, or at best, separation on a multitude of reservations, to an official policy of assimilation.

The students' assimilation into European-American civilization began immediately upon their arrival at Carlisle. They had to undergo the cutting of their hair, sacred in Native American culture. Males were made to dress in military uniforms, females in tight, restrictive dresses; gone were animal-skin clothing and moccasins. They ate the white man's food, went to the white man's churches, and spoke the white man's language. This process was meant to rid the children of their so-called heathen ways and was the epitome of "Killing the Indian, Saving the man." Sun Elk, a Taos Pueblo, said it well: "After a while, we also began to say Indians were bad. We laughed at our own people and their blankets and cooking pots and sacred societies and dances."

The school provided its students lessons in English and other academic subjects. It also taught the boys vocations, such as blacksmithing, tinning, harness making, shoemaking, carpentry, plumbing, bricklaying, telegraphy, printing, and tailoring. The girls learned cooking, sewing, and laundering.

The outcome of the Carlisle Indian Industrial School, however, is revealed in interesting, documented statistics: nearly twelve thousand students were enrolled, but only 758 students graduated, and at least 1,758 ran away from the school. By 1918 at the end of World War I, when it closed, Carlisle had produced a dual legacy: one, a major contribution to American sport's history and legend; the other, another failed American policy in the treatment of the Native Americans.

Dale E. Landon

See also: Assimilation; Boarding Schools; Haskell Institute; James Francis Thorpe.

FURTHER READING

Adams, David Wallace. *Education for Extinction: American Indians and the Boarding School Experience, 1875–1928*. Lawrence: University Press of Kansas, 1995.

Bloom, John. *To Show What an Indian Can Do: Athletics in Native American Boarding Schools*. Minneapolis: University of Minnesota Press, 2000.

Hall, Moss. *Go, Indians! Stories of the Great Indian Athletes of the Carlisle School*. Los Angeles: Ward Ritchie, 1971

Pratt, Richard Henry. *Battlefield and Classroom: Four Decades with the American Indian, 1867–1904*, ed. Robert M. Utley. New Haven: Yale University Press, 1964.

Phillip CASTILLO

Born 1973
Runner

Phillip Castillo, a native of Acoma Pueblo, New Mexico, was the first Native American to win an NCAA cross-country championship. Entering his first race at age six, Castillo demonstrated an innate talent for running. Although he ran in events organized by his pueblo throughout his childhood, it was not until he entered high school that he committed himself in earnest to extensive athletic training.

As a freshman at Grants High School, in Cibola County, New Mexico, he spent his time running on weekends and weeknights. His hard work paid off when he finished second in the state his sophomore year. In 1989, his senior year, he went to the Kinney Cross Country Championships, renamed the Foot Locker Cross Country Championship in 1993, where he placed eighth in the nation with a time of 15:19.1. While at Grants, he set a state and school running record.

Scouts from several colleges and universities offered him scholarships, but he chose Adams State College in Alamosa, CO, for its strong track and field program and for its proximity to home. While at Adams State, an NCAA Division II school, Castillo was an All-American in 1991, 1992, and 1993 and a 1992 national champion. In 1995 Castillo was honored as the Courageous Student Athlete (Male) at the Northeastern University Sport in Society Awards Banquet. Castillo's was ranked twenty-first in the nation when he placed sixth in the 1997 Las Vegas International Marathon and Half-Marathon with a time of 1:06:02 and a pace per mile of 5:02.

After receiving his master of science in physiology Castillo, joined the U.S. Army, where he was an 11M Bradley fighting-vehicle driver while participating in the U.S. Army World Class Athlete Program. The program identifies exceptional soldier-athletes and provides opportunities for them to compete in national and international events leading to qualifying trials for the U.S. Olympic team. Based in Colorado Springs, CO, and detailed for one year to Adams State University, Castillo trained for the 2000 Olympic trials, working on speed drills; he established in the Sacramento Marathon in December 1999 a qualifying time of 2:19.19, which ranked him twentieth in the nation. Earlier in October he had placed twenty-seventh at the Fifteenth Annual Army Ten Miler Qualification for the U.S. World Olympic team and achieved a time of 52:33.

Castillo ran in the Chicago Marathon and the U.S. National Cross-Country Championship in 2000 and also spent time at the San Diego Training Center. Since his time was under 2:20, his Olympic trail marathon expenses in May at Pittsburgh, PA, were covered. Only the top three runners of the trials made the 2000 Olympic team bound for Sydney, Australia. Castillo finished fifty-fifth in 2:34.22 and did not make the team.

He continued training for the next Olympic trials and used his athletic talent to inspire Native American youth by coordinating Wings of America, a summer outreach program based in Santa Fe, NM. In 2002 Castillo finished thirtieth in 47:30 in the USA Track & Field 15K championship. Based on results as of September 14, 2003, Castillo ranked eighty-fourth in the nation.

Rebecca Tolley-Stokes

FURTHER READING

"Foot Locker Cross Country Championships-1989 Finals-Boys" (footlockercc.com/history/1989boys.htm).

Mani, Thomas E. "'Muddy Sneakers' Set Record at Ten Miler" (www.dtic.mil/armylink/news/Oct1999/a1999101310mside.html).

Monastyrski, Jamie. "Pueblo Athlete Qualifies for Olympic Trials." *Gallup Independent* (members.home.nl/aeissing/00627.html), March 13, 2000.

Oberholser, Christian. "Acoma Runner Aims to Be Out in Front." *Gallup Independent* (www.gallupindependent.com/5-04-00.html), May 4, 2000.

———. "Acoma Runner Competes with World's Elite Class." *Gallup Independent* (www.gallupindependent.com/1-21-00.html), January 21, 2000.

★ ★ ★

Patty CATALANO

Born April 6, 1963
Runner

Micmac Patty Catalano (who also has competed as Patty Lyons and Patty Dillon) emerged as one of the elite women's distance runners of the late 1970s and early 1980s. Catalano, a native of Quincy, Massachusetts, started running to lose weight and improve her overall health. Not long after starting this fitness regimen, Catalano became enormously successful at several distances, none more so than the marathon. Amazingly, Catalano won the first marathon she ever entered, the Ocean State Marathon in Rhode Island in 1976. This was the start of an enormously successful tenure as a marathon runner.

Among Catalano's successes over the 26.2-mile course are her wins at the Montreal Marathon, the Houston Marathon, and four consecutive Honolulu Marathons, 1978–1981. In addition to the marathon, Catalano has won a number of events at other distances, including ten, twenty, and thirty-kilometer events. She set world records in distances ranging from five mile events to thirty kilometers.

Catalano's record at the Honolulu Marathon gives her the distinction as one of the most successful runners in that race's history, male or female. In her fourth and final Honolulu Marathon she set a course record in 2:33:24. Although she never won the Boston Marathon, the string of second-place finishes earned her respect among elite competitors. Moreover, in 1981 she set an American record in Boston at 2:27:51. Catalano's second-place finish at the New York City Marathon in 1980 is also noteworthy, as she was the first woman from the United States to run under two and a half hours in the event's history.

Like many distance runners, Catalano

logged over a hundred miles per week while training. This heavy mileage eventually took a toll on her, and by the early 1980s a series of injuries forced Catalano to withdraw from the sport. Despite her early departure from competitive distance running, Catalano left a mark on the sport and has been recognized for her efforts. She has been inducted into the Road Runners Club of America Hall of Fame and the Honolulu Marathon Hall of Fame.

Rita Liberti

FURTHER READING

Catalano, Patti, and Amby Burfoot. "The Comeback Trail." *Runner's World*, July 1984, 70–75.

Connelly, Colman. "Marathon Champ Honored." *Patriot Ledger* [Quincy, MA], December 14, 2002, 54.

Angela CHALMERS

Born 1963, Shiloh, Manitoba
Runner

Angela Chalmers, a Sioux middle-distance runner, was considered to be one of Canada's top runners of the 1990s. Born in Shilo, Manitoba, in 1963, Chalmers rose to

Many consider Angela Chalmers, here winning the 1,500 Meters at the IAAF Mobil Grand Prix, to be among Canada's greatest distance runners. *(AP/Wide World Photos)*

national athletic prominence beginning in the early 1980s, with silver-medal finishes in the 800 and 1,500-meter events at the 1981 Canada Summer Games in Thunder Bay, Ontario. Her strength as a runner continued to grow, leading to three second-place finishes from 1985 to 1988 at the Canadian Championships in the 1,500 meters. She first won that event at the Canadian Championships in 1989, with a time of 4:10.08. Chalmers was the first woman to win the 1,500 (4:08.41) and 3,000 (8:38.38) meter events at the 1990 Commonwealth Games in Auckland, New Zealand. In 1993 Chalmers had two first-place finishes in the 1,500 meters. Her first victory was at the Canadian Championships and the second at the World Championships in Stuttgart, Germany. Chalmers's record-breaking performance in the 3,000 meters at the 1994 Commonwealth Games in Victoria, British Columbia, was a Commonwealth Games record and a personal best (8:32.17). Chalmers's personal best in the 1,500 meters (4:01.61) at the 1994 Grand Prix Final in Paris, France, earned her the honor of being the first Canadian to win a Grand Prix event. In addition to medal wins at the Canadian Championships, Commonwealth Games, and the Grand Prix Final, Chalmers has earned medals at the Pan Am Games and the World Student Games.

Chalmers is a two-time Olympian. In 1988 she qualified for a spot on the Canadian Olympic team in the 1,500 and 3,000 meters. In 1992 at the Olympic Games in Barcelona, Spain Chalmers won a bronze medal in the 3,000 meters with a time of 8:47.22.

Angela Chalmers's accomplishments on the athletic field earned her a number of awards. She received the Manitoba Athlete of the Year Award in 1990 and in the same year was the recipient of the Phil Edwards Trophy, given to the outstanding Canadian track athlete. In 1994 she earned the Athletics Canada Athlete of the Year award. She is the 1995 National Aboriginal Achievement Award winner in the sports category. In recent years Chalmers has devoted her energy to working with the education ministry in curbing the high school dropout rate among aboriginal youth in British Columbia.

Rita Liberti

FURTHER READING

"Angela Chalmers." *Athletics Canada* (www.canoe.ca/AthcanAthletes/achalmers.html).

Gibb, Mike. "Special Delivery Countdown to Nagano: 17 Days to Go." *Toronto Sun,* January 21, 1998, 9.

Starkman, Randy. "Chalmers Fighting to Keep Youth on Track." *Toronto Sun,* July 5, 1997, B8.

Wilson "Buster" CHARLES

Born April 4, 1908
Decathlete, football player

An accomplished football and basketball player in both high school and college, Wilson "Buster" Charles (Oneida) gained his greatest fame as a member of the United States decathlon team at the 1932 Olympics, where he placed fourth. In 1972, his accomplishments as an all-around athlete earned him the distinction of being included in the first group of inductees to the American Indian Athletic Hall of Fame.

Son of Wilson Charles, Sr., a Carlisle Indian School graduate who had participated in both track and football in college and played professional baseball in the early 1900s, Wilson attended high school at Flandreau Indian School in South Dakota, where his father served as boys' adviser. While in South Dakota, Charles won All-

Wilson "Buster" Charles was a multi-sport talent, excelling at football and track and field. *(AP/Wide World Photos)*

State honors in football in 1926 and 1927, and won the state championship in the high jump and long jump in 1927. In the summer of 1927, he moved to Lawrence, Kansas, to attend the Haskell Institute, where he gained notoriety in all three of his high school sports. At six feet, two inches and two hundred pounds, and wearing number fifty-two, Charles played as fullback while also handling kicking and punting duties.

His head football coach later referred to him as a "triple threat"—passer, runner, and kicker. He also credited Charles as being one of the greatest punters and all-around athletes ever to play for Haskell. Charles helped lead the football team to ten victories in eleven games in 1930, its only loss being to the local University of Kansas team. He also played center for the Haskell basketball team, but his greatest love was the decathlon, a track and field event composed of the high jump, long jump, hundred-meter dash, four-hundred-meter dash, 110-meter high hurdles, 1,500 meter run, shotput, discus, javelin, and pole vault. In 1930, competing in the decathlon, Charles won the Kansas Relays championship and the National Amateur Athletic Union title.

At the Olympic Games in Los Angeles, Charles led the field after the first five events in the decathlon competition, but his weaker performance in the pole vault cost him a medal, leaving him in fourth place with a score of 7,985 points. The gold medal went to his American teammate Jim Bausch, who scored 8,462 points. Bausch competed for the University of Kansas, also located in Lawrence.

After graduating from Haskell in 1932, Charles continued on to the University of New Mexico in Albuquerque, where he met his wife, Nola. In 1938 the couple moved to Phoenix, Arizona, where Wilson took a job in an engineering firm. They later retired to Camp Verde, Arizona. In addition to his selection to the American Indian Athletic Hall of Fame, Charles was honored in a 1972 Haskell reunion, named as an honorary citizen of Lawrence by the city commission, and was inducted into both the Howard Wood Dakota Relays Hall of Fame and the South Dakota Hall of Fame.

Wade Davies

FURTHER READING

Charles, Wilson "Buster" file. *American Indian Athletic Hall of Fame Collection, Cultural Center and Museum.* Lawrence, KS: Haskell Indian Nations University.

Oxendine, Joseph B. *American Indian Sports Heritage.* Lincoln: University of Nebraska Press, 1995.

Oscar CHARLESTON

Born October 14, 1896, Indianapolis, Indiana
Died October 6, 1954, Philadelphia, Pennsylvania
Baseball player

Oscar Charleston, a heavy-hitting center fielder in baseball's Negro leagues, was considered by many to be the greatest baseball player of all time. Combining the explosive speed of Ty Cobb, the power of Babe Ruth, and the defensive ability of Willie Mays, Charleston's diverse skills made him perhaps the most well-rounded player in the history of the game. An imposing, barrel-chested man with a quick temper, he played nearly three decades with the Indianapolis ABCs, Harrisburg Giants, and other Negro league teams.

Charleston was born in 1896, in Indianapolis, to Tom Charleston, a Sioux construction worker, and Mary Thomas, an African-American woman. At age fifteen he ran away from home and lied about his age to join the army. He was stationed in the Philippines, where he ran track and first began to play baseball seriously. After his discharge in 1915 he joined the Indianapolis ABCs, a renowned African-American team in his hometown. He soon became famous for both his fiery disposition and his exceptional play. Charleston was also an outstanding fielder, base runner, and bunter.

In 1924 Charleston became player-manager of the Harrisburg Giants in the newly formed Eastern Colored League. He enjoyed several of his finest seasons there, batting a reported .411 in 1924 and .445 in 1925, while leading the league in homers both years. Charleston continued his outstanding play during brief stints with the Hilldale Daisies and Homestead Grays before joining the Pittsburgh Crawfords as player-manager in 1932. The Crawfords, featuring five future Hall of Famers, won pennants in 1935 and 1936. Charleston had moved to first base in 1930 to make room for younger, faster outfielders but was still as dangerous a hitter as ever, posting batting averages ranging from .288 to .450 during the 1930s.

Early in his career Charleston had a quick temper and was known for getting into fights on and off the field. He was "a devilish type of guy," one teammate later remembered. "Always grinning, always playing pranks, always jolly. But deep back inside you could see he was a cold-blooded son of a gun." On October 29, 1915, during his rookie season, he was arrested for punching an umpire after a disputed call during an exhibition game against a major league All-Star team. Although always a fierce competitor on the field, his temper mellowed in his later years; as manager of the Pittsburgh Crawfords and Philadelphia Stars, he served as a mentor to younger players.

In 195 recorded at-bats against major league pitching during his career, Charleston batted .318, with a .569 slugging percentage. Over nine winter seasons in Cuba, he posted a cumulative .361 average. On October 5, 1954, shortly after managing the Indianapolis Clowns—a team featuring two female players—to the Negro American League title, Charleston suffered a heart attack. He died in Philadelphia the next day, having spent forty of his fifty-seven years in professional baseball. Many honors were accorded him in the years after his death, including induction into the Baseball Hall of Fame in 1976. In 2001, renowned baseball historian Bill James named Charleston as the fourth-greatest player in baseball history, behind only Babe Ruth, Honus Wagner, and Willie Mays.

Eric Enders

FURTHER READING

Clark, Dick, and John B. Holway. "Charleston No. 1 Star of 1921 Negro League." *Baseball Research Journal* (1985): 63–70.

Holway, John. *Blackball Stars.* New York: Carroll and Graf, 1988.
Lester, Larry, and Dick Clark, eds. *The Negro Leagues Book.* Cleveland, OH: Society for American Baseball Research, 1994.
Research Files. National Baseball Hall of Fame Library, Cooperstown, NY.
Riley, James. *The Biographical Encyclopedia of the Negro Baseball Leagues.* New York: Carroll and Graf, 1994.

Virgil Earl CHEEVES

Born February 12, 1901
Died May 5, 1979
Baseball player

Cheeves, a Cherokee pitcher, played professional baseball in the decade after the First World War.

Cheeves began playing baseball at Bowie High School in Dallas, Texas, before pitching for the Eastland club of the West Texas League in 1919. In the latter part of the 1920 season, the Chicago Cubs recruited Cheeves; he pitched five games for them that year, with no decisions.

In 1921, when Cheeves was barely twenty, the April 15 *Spring Training News* reported that Cheeves "has good pitching sessions." Cheeves was right-handed, six feet tall, and weighed 185 pounds. During 1921–1922, Virgil "Chief" Cheeves enjoyed the nickname of "Giant Killer," because he beat the championship New York Giants the first six times that he pitched against them. In 1923, however, he hurt his arm and was traded to the Cleveland Indians and then to the St. Louis Cardinals, for whom he played the role of middle reliever in the minor league system. In 1927, Cheeves returned to the major leagues, pitching for the New York Giants, posting twenty-six wins and twenty-seven losses.

After retiring from baseball, Cheeves worked in the painting and construction industry and in 1963, he married his second wife, Allie Dillard Cox.

Kelly Boyer Sargent

FURTHER READING

The Ballplayers Mini-Biography (www.baseball-almanac.com/players/player.php?p?heevvi01).
Baseball-Reference.com (www.baseball-reference.com/c/cheevvi01.shtml).

CHEMAWA INDIAN SCHOOL

Chemawa Indian School is one of only five boarding schools still managed by the Bureau of Indian Affairs for the education of Native American youth. This first off-reservation boarding school on the West Coast was approved for construction in 1879, simultaneously with its eastern counterpart, Carlisle. Among Indian people and Native Alaskans, family traditions of attending Chemawa have been a significant factor in its continuing existence. Its sports program has been one of the most important aspects of the curriculum. The school originated at Forest Grove Oregon as the Forest Grove Indian and Industrial Training School. Moved in 1885 to a site on the railroad in what is now Keizer, Oregon, it became the Salem Indian Training School and was briefly known as the Harrison Institute. By the early twentieth century, it had become known simply as Chemawa.

The first superintendent, Lt. Melville C. Wilkenson, had been, like his contemporary Richard Henry Pratt, of Carlisle fame, an officer in the Union army. Ironically, Wilkinson had been the aide-de-camp for Gen. Oliver Otis Howard during the campaign against Chief Joseph only three years before the school was established. The school was initially designed as an "industrial training school" with few academic classes and a singular focus on assimilating and acculturating Indian students and preparing them for work in factories and farms. Girls were trained as domestics, either to return to the reservation carrying their newly instilled Euro-American values or to work for white families as maidservants and babysitters.

The early curriculum included a form of athletic training through calisthenics and marching drills for all students. The student newspaper at Forest Grove in April 1884 reported the formation of the school's first baseball team, composed of the young men who were the first "scholars" selected from northwestern tribes to attend and help construct the school itself. Forbidden to speak anything but English, they were forced to abide by the standards and mores of white society. But working and studying, living and playing with members of other tribes and Alaskan villages had a consequence unanticipated by the federal authorities. It provided the context in which Indian students would develop what would be known as the "Pan-Indian" movement, and that movement in many ways ensured the survival of Chemawa.

In the first years, athletic activities were not part of the curriculum, but the students used the dairy barn as a gymnasium and played their own ball games out in the pastures—in their rare free time. This type of activity was not encouraged by some in the Indian Service, as illustrated by the 1889 report to the secretary of the interior from M.M. Waldron, physician at the Hampton Institute: "Their violent games and races task their strength to the utmost for the time, but often at the expense of some vital organs. The result is protracted inactivity and general demoralization. Civilization is gradually correcting all this, and better physical development will be the result."

Sports at Chemawa were not discouraged, however, and at the Commencement Field Day in June 1889 the violent games and races bemoaned at Hampton were celebrated with a competition in the 220-yard run, 220-yard hurdles, 100-yard dash, running broad jump, mile walk, high jump, 440-yard run, shotput, half-mile run, mile run, and pole vault. That year the baseball team had beaten McMinnville College and the Multnomah Club, and the Portland YMCA had bowed to both Chemawa's football and baseball teams.

The success of athletic programs, coupled with an increased concern over the death rate of Indian students at all of the boarding schools, made it clear that assimilation via institutionalization was not quite as healthful as first supposed. In 1890, new "Rules for Indian Schools" were published in the Annual Report of the Commissioner of Indian Affairs, requiring special hours for recreation and outdoor sports. Students were to be taught the "games enjoyed by white youth, such as baseball, hopscotch, croquet, marbles, bean bags, dominoes, checkers, logomachy, and other word and letter games." The girls were relegated to knitting, netting, crocheting, and embroidery.

By 1896, Chemawa boys were playing football, baseball, basketball, and tennis, while the girls had two basketball teams, which had won several victories. At that time, the students were able to complete only the eighth grade, so it is important to note that the Indian athletes were not children but young adults—at no disadvantage in a game with the local college teams. The Indian school was beginning to establish its reputation and Reuben Sanders (known as "Chemawa's Jim Thorpe") was in his heyday.

At the turn of the century, baseball, football, basketball, tennis, and croquet were regular parts of the school's extracurricular activities. The parallel development of the school band reinforced enthusiasm for the athletic program; fans could enjoy both competition and musical entertainment. A proliferation of clubs and organizations filled the students' time. With names like the Excelsior and Reliance Literary Societies for the boys and Winona and Nonpareil Literary Societies for the girls, there was no doubt as to the ongoing commitment of the government to acculturation of Indian students.

Between 1897 and 1907, the Indians, Chemawa's football team (for a time known as the Redskins, now the Braves), played the University of Oregon five times, Oregon State University four times, and the University of Washington two times. In addition, in 1903 the team traveled to San Francisco to play the University of California, Berkeley, and Stanford University. For two years, from 1916 to 1917, William J. Warner, brother of Carlisle Indian School's Glenn "Pop" Warner, coached the team, and in 1921 Chemawa played Oregon State University again.

Complementing these sports was the Salem to Portland Relay Race held every May between 1907 and 1912. Covering a distance of fifty-one miles, this marathon event pitted ten of the fastest Chemawa runners against ten for the Portland YMCA Attracting statewide interest, the inaugural event was won by Chemawa in a time of just under five and one-half hours. Of the six contests, Chemawa won four and the YMCA two. The victorious team received a banquet, a silver cup, gold medals, and a pie to eat that measured two feet wide and five feet long. Chemawa's early athletic reputation was largely driven by the efforts of Reuben Sanders, who had achieved national fame in his own right. Returning from Sherman Institute in 1911, he served as Chemawa's coach for the next thirty years. During his tenure, wrestling, the javelin, discus throw, and cross-country were added to the school's athletic program.

By the 1930s, the Indian education system was in the process of radical overhaul, as was the tradition of a "military" system for Indian schools. Students in uniforms, class battalions and officers, and marching to and from the classes became a thing of the past. Where trophies for the winners of competitive drills once stood on display, new gold statues glittered, honoring the school's athletic excellence. In 1944, an article in the Oregon Statesman reported on Chemawa's football team, leaders of the western sector of the state in class B competition. Then coached by "Chief" Tommy Thompson and Rube Sanders, the team had already played seven of their ten games that year against class A competition in the Duration League and was considered one of the top contenders for the state championship.

Changes on both a local and national level would impact Chemawa and its sports. Rampant diseases like the Spanish flu and tuberculosis, two world wars, and other conflicts removed many potential sports heroes from the rosters. Shifts in the policies of the Bureau of Indian Affairs (BIA) emphasized the construction of schools on the reservations and placement of Indian students in local schools. After 1900, off-reservation boarding schools began to disappear, and Chemawa made the BIA "hit list" on many occasions. But alumni from tribes all over the nation protested, and the school managed to not only survive but reinforce its position as a unique and necessary institution. Its location and ability to serve students with few educational options from Alaska and the Southwest made the school an imperative. Through all the changes, the sports program has remained and is supported to this day as one of the most important aspects of life at Chemawa. The athletic ability of Chemawa's teams continues to be a proud tradition.

SuAnn M. Reddick and Cary C. Collins

See also: Assimilation; Boarding Schools; Carlisle Indian Industrial School; Haskell Institute.

FURTHER READING

Collins, Cary C. "The Broken Crucible of Assimilation: Forest Grove Indian School and the Origins of Off-Reservation Boarding-School Education in the West." *Oregon Historical Quarterly* 101:4 (Fall 2000): 444–65.

———. "Oregon's Carlisle: A History of Chemawa Indian School." *Columbia: The Magazine of Northwest History* 12:2 (Summer 1998): 6–10.

Lomawaima, K. Tsianina. *They Called it Prairie Light: The Story of Chilocco Indian School*. Lincoln: University of Nebraska Press, 1994.

Reddick, SuAnn M. "The Evolution of Chemawa Indian School: From Red River fo Salem, 1825–1885." *Oregon Historical Quarterly* 101:4 (Fall 2000): 444–65.

Trennert, Robert A. *The Phoenix Indian School: Forced Assimilation in Arizona, 1891–1935*. Norman: University of Oklahoma Press, 1988.

CHEROKEE MARBLES

Just like the Cherokee culture, the game of marbles is ancient. The game, which may date back to A.D. 800 or even older, according to tribal tradition, relies on skill and strategy to negotiate a five-hole course with a ball that is referred to as a marble.

In ancient times, marbles were made of flint stone. After the flint stone was chipped away with a rock or a chipping tool to form a nearly round ball, the rough edges were smoothed out with a big sandstone on a stick. The marble was rolled against the sandstone until it was about the size of a billiard ball. Today, Cherokee players in northeastern Oklahoma use a white billiard ball.

To form the playing field, players make five indentations or holes in a row, thirty yards apart. The fifth hole is made thirty yards to the left or right of the fourth hole. Players attempt to roll their marble into the first hole, then the second, progressing to the fifth hole and then back to the first hole. A player may knock a competitor's marble away from holes by striking it with their marble. The first team to have all its players complete the course wins.

There are two sets of rules for Cherokee marbles, tournament and "cheaters." With tournament rules, the teams comprise three starters and one alternate. In cheaters' rules, a team may have more or less than four players. On the first hole in a tournament game, there is no hitting. But on the second hole, a player can hit an opponent's ball twice per hole per opponent. In a tournament, the marble must be inside the circle around a hole to drop it in. In cheaters, which has more lenient rules, players can step over the marble line to throw a marble. In a tournament game, players stay behind a stick that they carry.

A Cherokee marbles annual tournament is played on the tribal headquarters grounds near Tahlequah, Oklahoma, at the Cherokee Nation of Oklahoma's annual holiday during each Labor Day weekend. During the year, lights are set up on the tournament grounds for play each Thursday evening, weather permitting.

Royse Parr

FURTHER READING

Snell, Travis. "Cherokee Marbles." *Cherokee Phoenix and Indian Advocate* 26:1 (Winter 2002).

CHIEF BIG HAWK (Kootahwe Cootsoolehoo La Shar)

Born c. 1850
Died Unknown
Scout, runner

Chief Big Hawk, while serving with the Pawnee Scouts in 1876, was the first man to run a mile in under four minutes.

Well known for their running abilities, the Pawnee have a long history of cross-country athleticism. The Pawnee, like other Native nations, relied upon runners to carry messages from village to village.

Big Hawk's phenomenal running times were frequently disbelieved, though they were documented twice. Big Hawk ran approximately 120 miles between the Pawnee Agency, modern Pawnee County, OK, and the Wichita Agency near Fort Cobb, modern Caddo County, OK, in under twenty-four hours; his critics challenged his feat as an impossibility. To settle the matter, the Wichita chief (unidentified in the record but most likely Tawakoni Jim) accompanied him on the trip back, making arrangements for a fresh horse for himself at the sixty-mile mark.

However, the chief's horse gave out before reaching that point, forcing him to stop. Big Hawk continued running. Upon arriving at the Pawnee village just before sunrise, the Wichita chief found Big Hawk sound asleep less than twenty-four hours after their departure. Big Hawk reached the Pawnee village just around midnight, thus completing the 120-mile run in approximately twenty hours.

In fact, Big Hawk ran a mile twice in under four minutes at Fort Sidney, NE. Accounts state that a Captain North set up a mile-long course for Big Hawk to run. Using two stopwatches, observers timed him as he ran the route. On his first attempt, Big Hawk reached the halfway point in the course at two minutes. In disbelief, North had Big Hawk run the course a second time to confirm that the time was correct, and measured the course a second time to make certain that the distance was in fact one mile. On Big Hawk's second attempt, he reached the halfway mark at 1:58 minutes and finished at 3:58. The fastest recorded sub-four-minute mile was recorded at 4:49 almost eighty years later in 1954, when Roger Bannister received the credit. Big Hawk was never credited for his remarkable record-setting time.

Rebecca Tolley-Stokes

See also: Running.

FURTHER READING

Gilbert, Bill. "Big Hawk Chief, A Pawnee Runner." *American West* 21:4 (July/August 1984): 36–38.
Johansen, Bruce, ed. *Encyclopedia of Native American Biography: Six Hundred Life Stories of Important People from Powhatan to Wilma Mankiller.* New York: Henry Holt, 1997.
Oxendine, Joseph B. *American Indian Sports Heritage.* Champaign, IL: Human Kinetics Books, 1988.
Waldman, Carl. *Who Was Who in Native American History: Indians and Non-Indians from Early Contacts Through 1900.* New York: Facts On File, 1990.

CHUNKEY

Chunkey was an important sport of dexterity and skill that performed an integral role in the social, political, religious, and economic activities of most of the indigenous nations in what is today the southeastern United States.

Rules

As with the hoop-and-pole game played by a majority of the Native American groups north of Mexico, the indigenous people of the Southeast played chunkey for centuries prior to the arrival of Europeans in the region and continued to play it well into the nineteenth century. Usually, two males from the same town opposed each other in a game of chunkey. A single game contained multiple rounds, and as each round began one of the two contestants rolled a circular, stone disk along its edge across a flattened courtyard. When the chunkey stone reached a certain point in the courtyard, each of the players ran after it and tried to slide or throw his respective pole along the ground in an effort to have it be the closest to the stone when it finally came to a rest. Usually made of agate or quartzite, the chunkey stone was typically circular in shape (about three to five inches in diameter), convex on one side, flat on the other side, and had depressions near, or a hole through, the center on both sides for thumb and finger grips. The poles were anywhere from eight to fifteen feet long and two to three inches in diameter. Often the players smeared bear grease along the length of the poles to make them slide more smoothly across the ground. Scoring usually consisted of one point given to the one with his pole closest to the chunkey stone when it stopped rolling or two points if his pole actually touched the stone when it came to a rest. The player that won the round would then throw the chunkey stone in the next round. If both poles were the same distance from the chunkey stone, then no one scored, and the hurler of the stone from the previous round threw the chunkey stone in the next round. The first player to reach eleven points won. Playing strategy consisted of either trying to knock the opponent's pole away from the chunkey stone or simply sliding the pole as close to the stone as possible.

Playing Area

Chunkey was played in the center of town in the main plaza, where all the important religious and political activities for the town took place. This alone reflects the importance of chunkey to the indigenous peoples of the Southeast United States. In fact since the playing of chunkey was so closely associated with the central courtyard of the town, this area was often referred to as the *ìchunk yardî*, despite the fact that many other important activities took place there as well.

These plazas were rectangular in shape and ran between two hundred and nine hundred feet in length, with a proportionate width. Perfectly level throughout its length and width except for a low mound in the middle, this plaza sat two or three feet below one or two rows of terraced embankments created from the excess earth removed from the plaza in the process of making it level. These embankments served as seats for spectators and ran along the length of the chunk yard. The surface of the plaza was packed clay covered with a thin layer of sand.

On the low mound in the center of the plaza stood a pole thirty or forty feet high made of pine. Some type of effigy (usually a bird) or a platform that represented a nest topped the pole. This pole and its top piece sometimes functioned as a target for archery or musket competitions. It also was an integral part of one version of the Southeastern racquetball or stickball game, as well as the centerpiece for many important dances and ceremonies. The winter council house anchored one end of the chunk yard, and a public square with awnings made of saplings and grass mats lay at the other end. This second area served as the spot for town and council meetings in the warmer months. At one end of the chunk yard, a twelve-foot pole stood near each of the corners. These were slave poles, on which prisoners captured in war were tortured and killed. Each was adorned with trophy scalps from past victories. Clearly, the chunk yard occupied the busiest section of the village. The playing of chunkey in the heart of the town, where the most important community activities took place, demonstrates the importance of the game to the town and that it was interwoven into the social, religious, and political fabric of the community.

Purpose of Chunkey

Chunkey was played from as early as A.D. 900 into the nineteenth century. Throughout the Mississippian era (A.D. 900–1600) of indigenous cultural development, chunkey seems to have been a vital part of the religious, political, and economic activities of native towns throughout the Southeast. Chunkey had its own set of ceremonial rites, to be performed by participants and spectators alike to ensure the victory of a particular contestant. The game was so important to native communities that chunkey stones were considered community property; they were kept in the council house and were handed down from one generation to the next. Even though chunkey was a game of skill and dexterity, the native people of the Southeast also saw it as a contest of spiritual power; the winner garnered new power and prestige. Thus by successfully playing chunkey, an individual male improved his political standing in the community, having demonstrated that he had spiritual knowledge and power. This in turn might help the individual to become a war leader or even a spiritual leader.

Because participants and spectators bet heavily upon the outcome of these matches, chunkey games also served as economic redistribution mechanisms. Sometimes participants bet everything they owned upon a contest, but the community generally discouraged this. In the end, people who had more tended to gamble more, and this along with the frequency of matches played helped guarantee that the wealth of the community was more or less evenly distributed.

Dixie Ray Haggard

See also: Hoop and Pole.

FURTHER READING

Cochran, Catherine. "Traditional Adult Cherokee Games." *Journal of Cherokee Studies* 13 (1992): 12–4.
Culin, Stewart. *Games of the North American Indians*. Vol. II, *Games of Skill*. Lincoln: University of Nebraska Press, 1992.

Jack "Sammy" CLAPHAN

Born October 10, 1956
Died November 26, 2001
Football player

Jack "Sammy" Claphan devoted much of his life to football. He excelled on the gridiron in high school, college, and later in the NFL, before returning to coach his high school alma mater.

Claphan's first organized football experience came in 1970, when he tried out for the Stilwell High School football team as a walk-on. He surprised coaches and players by making the varsity team his freshman year. By his sophomore year, Claphan, wearing jersey number seventy, was playing on both ends of the ball, excelling at both offense and defense. Best known as a football player, he was a versatile athlete who also wrestled, played basketball, and ran on the track teams, lettering on each team. His football jersey was retired by his high school in the mid-1980s.

After graduating from Stilwell High School, Claphan received a football scholarship and continued his football career at the University of Oklahoma, playing offensive guard. As a red-shirted freshman, wearing jersey number sixty-three, he played in only a few games. However, Claphan was starting by his junior year, and he lettered in football in 1976, 1977, and 1978. By his senior year in college he and his linemates had become the most dominant line in the country, clearing the way for a teammate to win the coveted Heisman trophy.

In 1979, Claphan graduated from the University of Oklahoma and entered the professional football draft. He was drafted in the second round (forty-seventh overall pick) by the Cleveland Browns, but he never made the roster, as a result of a preseason injury. The next season (1980) he was picked up by the San Diego Chargers, and the next year he made the transition from the practice squad to the official team roster. In his first year starting, Claphan played mostly as a special teams lineman. During the rest of his time with the Chargers, Claphan started at both the left guard and left tackle positions, starting at left tackle until he left the team in 1987.

Claphan's excellence in football at both the high school and college levels had brought him more than letters. In April of 1994, Claphan was inducted into the American Indian Athletic Hall of Fame in Albuquerque, New Mexico.

After his tenure with the Chargers, Claphan returned home to take up a coaching career, eventually coaching at Stillwell High School for two years. Claphan's life extended beyond football, as after returning home from the San Diego Chargers, he became active in both his community and his tribe. Politically, he participated in the election of Cherokee chief in May of 1999, receiving over 500 votes, and he also ran for a county council seat.

Claphan then taught special education in the Stilwell school system until November 26, 2001, when he died of a heart attack. He left a wife (Linda), a daughter, Amber (who is continuing her father's legacy by attending Oklahoma University), and a son, Erik. He was also survived by his mother, two sisters, two brothers, and four nieces.

Beth Pamela Jacobson

Jay Justin "Nig" CLARKE

Born December 15, 1882, Amherstburg, Ontario
Died June 15, 1949, River Rouge, Michigan
Baseball player

Jay Clarke was a light-hitting catcher in the early twentieth century for the Cleveland Indians, Detroit Tigers, St. Louis Browns, Philadelphia Phillies, and Pittsburgh Pirates. Born in 1882, in Amherstburg, Ontario, Clarke was one of the first baseball players from Canada to play in the major leagues. In nine seasons, he hit a mediocre .254, with six home runs and 127 RBIs. The main reason he remained in the big leagues was his defensive abilities, though they were hampered by arm problems, and the fact that he was a switch-hitter whom managers could use against tough right-handers. Ultimately he was sent to the minor leagues, where he stayed for seven years during the prime of his career.

At nineteen, Clarke first gained national attention while playing for Corsicana in the Texas League against Texarkana on June 15, 1902. Clarke's team trounced Texarkana 51–3. While it was alleged that the pitcher gave him pitches to hit, Clarke led his team in hitting, going eight for eight, and collected sixteen RBIs and thirty-two total bases. These were all organized-baseball records at the time.

Arriving in the major leagues three years later, at the age of twenty-two, the five-foot eight-inch, 165-pound Clarke caught thirty-seven games for the Cleveland Indians, hitting an embarrassing .202 in 114 at bats. When the Detroit Tigers needed a catcher in August, Cleveland agreed to loan the young catcher to them. Joining the Tigers on August 1, he went three for seven for them before being returned to the Indians after ten days. This was not an uncommon practice in the first decade of the twentieth century; two other catchers were involved in similar transactions that year. Clarke spent the next two seasons mainly as a reserve, seeing action in forty-five and fifty-seven games, respectively, in the next two seasons. He became a starter in 1907 and hit a respectable .269, getting three out of his six major league home runs. Released in 1911, he began a new career as a minor league journeyman. To the shock of many observers, he resurfaced in the major leagues in 1919 for the Philadelphia Phillies as a reserve catcher. He hit a decent .242 in sixty-two at-bats. He tried to continue his comeback the next season for the Pittsburgh Pirates but was released after just three games. Clarke died on June 15, 1949, in River Rouge, Michigan.

T. Jason Soderstrum

FURTHER READING

Blaha, Thomas. "Canada's Contribution to America's National Pastime." *Saskatchewan Historical Baseball Review.* Battleford: Saskatchewan Baseball Hall of Fame and Museum, 1993.

Davis, Mac. *The Lore and Legends of Baseball.* New York: Lantern, 1953.

Smith, Red. "Views of Sport." *Pacific Stars & Stripes,* May 15, 1953, 13.

"Some Happenings at Mobile." *Sporting News,* June 29, 1916, 1.

James E. "Jimmy" CLAXTON

Born 1892, Vancouver Island, Claxton
Died 1970
Baseball player

Jimmy Claxton was a victim of organized baseball's strict racial policies in the early twentieth century. The son of an English-Irish mother and a father of African-American, Native American, and French descent, Claxton played briefly for the Oakland Oaks of the Pacific Coast League in 1916 before being released due to his African-American heritage.

Born on Vancouver Island, Claxton moved with his family to Washington State before his first birthday. He reportedly began playing competitive baseball for an amateur team in Roslyn, Washington, in 1912 when he was only thirteen years old. Claxton pitched for local and semipro teams throughout his teen years and joined the all-black Oakland Giants in 1916.

Claxton made his organized baseball debut on May 28, 1916, as the starting pitcher for the Oaks in the first game of a doubleheader against the Los Angeles Angels. He gave up four runs in his only two innings of work, but he came back to get the last out of the second game.

Claxton's tenure with the Oaks was short-lived, however, and he never appeared in another PCL game. Soon after joining the team, someone informed Oakland's management of Claxton's African-American background, and the Oaks promptly released him on June 2. The manager of Oakland excused his team's transgression against the unwritten bar on

African Americans by claiming that Claxton had shown him a notarized affidavit affirming his Native American heritage.

After his short stint in the Pacific Coast League, Claxton played baseball throughout the country, mostly for black teams, including the Chicago Union Giants and, in 1932, the Cuban Stars of the East-West (Negro) League. He also played several seasons for an all Native American barnstorming team, the Nebraska Indians.

Gregory Bond

FURTHER READING

Walton, Dan. "Sports-Log." *Tacoma News-Tribune*, May 17, 1964.

Weiss, William J. "The First Negro in 20th Century O.B." *Baseball Research Journal* (1979): 31–35.

Harold COLLINS

Born May 25, 1957, Shannon, North Carolina
Weightlifter

Harold Dean Collins, a Lumbee known as "Chief Iron Bear," has amassed a remarkable record for powerlifting and feats of strength, five of which are recorded as Guinness World Records.

Collins was born in 1957, in Shannon, North Carolina. His parents are Redell and Evelyn Collins. He owns and operates the Pembroke Power House Gym in the Lumbee population center of Pembroke, North Carolina. At age five, he set himself the goal of becoming the world's strongest Native American. His numerous strength and powerlifting accomplishments include the following: setting a still-unbroken North Carolina bench-press record of 633 pounds in the super-heavyweight class in 1984; U.S. Powerlifting Federation champion in the 125-plus kilo class in 1991, 1992, and 1993; gold medalist in the World Bench Press competition in 1991 and 1992; first runner-up in the U.S. Strongest Man contest in 1997; and sixth place in 1993 and tenth place in 1997 in the World's Strongest Man contest (which consists of eight events).

Collins has long been committed to social and charitable causes, especially working with youth. As part of his "Say No to Drugs" campaign, he has visited schools and hospitals in North Carolina, Texas, Arizona, the Dakotas, Canada, Saudi Arabia, Scotland, Lithuania, and Russia. In 1996 he raised $100,000 for the March of Dimes by bench-pressing $10,000 in dimes for ten repetitions. The televised event was held in Hollywood Park, Los Angeles. In 2002, he held sports camps for youth in his home region of Robeson County, North Carolina. He also established the Native American Strength Association. One goal of the organization is to teach Native American youth correct weightlifting techniques and train them to compete effectively in strength contests. To that end, he brought Native American youth from throughout the United States and Canada to his gym to learn training techniques from him.

The Lumbee tribe has also been a significant focus for Collins. In 2000 he was a candidate for tribal chairperson. He has been deeply involved in planning and fund-raising for the North Carolina Indian Cultural Center, raising $10,000 by auctioning commemorative Lumbee high-top sneakers that he designed in partnership with the local Converse factory.

Glenn Ellen Starr Stilling

FURTHER READING

"'Chief Iron Bear' Harold Collins Files for Tribal Chairman." *Carolina Indian Voice*, September 28, 2000, 1.

Fox, Geoff. "Collins Wins Toe-to-Tire War." *Robesonian*, April 23, 2002.

Steve COLLINS

Born May 13, 1964
Ski jumper

Steve Collins, an Ojibwa, was one of the most talented ski jumpers in the world during the 1980s. Collins grew up on the Fort William Reserve near Thunder Bay, Ontario. A natural talent, he leaped from obscurity as the youngest competitor ever to compete in a world event in Finland. By the age of fifteen, he was already a celebrity in Europe, having won three major ski jumping events.

Steve Collins, along with Horst Bulau and Ron Richards, made Canada a formidable jumping power during the 1980s with several international ski jumping victories. In 1980, at the age of sixteen, Collins finished ninth in the ninety-meter jump at the Winter Olympics in Lake Placid. This was the highest finish ever for a Canadian ski jumper at the Olympics. He was also top Canadian in the seventy meter, finishing in twenty-eighth place. That same year, Collins reached the height of his career when he won the Czechoslovakia ski flying championship, as well as a World Cup competition and the World Junior Championship, leaping to a record 124 meters at Lahti, Finland. In Canada he was presented with the Tom Longboat Award, given to the outstanding Aboriginal athlete, and the Viscount Alexander Trophy, presented annually to Canada's outstanding junior male athlete of the year. The next year he won the U.S. National Junior Ski Jumping Championship and set a world-best ninety-meter ski-jumping mark, leaping 128.5 meters in Thunder Bay.

Traveling on the World Cup circuit, being a celebrity at such a young age, and missing his familiar surroundings at the Fort William reserve was too much for a young man to handle. Collins soon lost his competitive edge and slipped down to ninety-sixth place in World Cup Standings. Prior to the 1984 Olympics in Sarajevo, Collins was sent to an addiction rehabilitation center. At the Olympics, he was the top Canadian in the seventy-meter event, placing twenty-fifth, the best ever finish by a Canadian in the seventy-meter competition. He also finished in thirty-sixth position in the ninety-meter jump. The next year Collins registered five top-ten finishes on the World Cup Circuit. In the 1988 Winter Olympics, in his home country of Canada, Collins finished his career with a thirteenth-place finish in the seventy-meter, improving upon his best-ever finish by a Canadian set in the previous Olympic Games. He placed thirty-sixth in the ninety-meter jump and led Canada to a ninth-place finish in the team jumping event. He retired from the World Cup circuit in 1991 and was inducted into the Canadian Ski Hall of Fame in 1995.

Edward W. Hathaway

FURTHER READING

Bryden, Wendy. *Canada at the Olympic Winter Games.* Edmonton, ALTA: Hurtig, 1987.

COMPETITION POWWOWS

Residential schools and bureaucratic policy served to reduce and even eliminate several Aboriginal means of cultural expression, including powwows, during the late 1800s and early 1900s. However, change occurred during the 1960s civil rights era, when the Red Power movement helped to bring Native oppression into the spotlight, resulting in renewed interest in Aboriginal culture and the development of competition powwows. As a "traditional" event,

Powwows have become important expressions of heritage and athleticism. *(AP/Wide World Photos)*

the powwow has been conceptualized as a static residual practice, a remnant of the past. However, an increase in participation in powwows as well as concomitant changes that have occurred in the past fifty years have forced the reconceptualization of powwows in a way that allows contemporary competition powwows to be included in the ever-expanding category of "sport."

Origins

Plains Indian grass dancing is generally credited as the origin of the powwow, though powwows are now practiced by Aboriginal and American Indians across Turtle Island (North America). Roberts (1992) found that American Indians in Oklahoma and Nebraska spread powwow dancing to the Sioux, who in turn passed it along to other tribes. The word "powwow"—often written "pow wow" or "pow-wow"—is a European adaptation of the Algonquin term *Pau Wau,* which referred to "the tribal medicine men and spiritual leaders." This term was later expanded to mean "the entire gathering of tribes."

Just as the meaning of the "powwow" has expanded and changed, so too has the practice of powwow dancing. Over the past century, the traditional powwow has represented different things to different groups of people. Early federal bureaucrats and religious figures often viewed powwows as the devil's work. More informed individuals view powwows in other ways. Some see them as giant family unions and cultural celebrations, revolving around feasting and dancing; others take a more sociological approach, viewing powwows, such as those in the 1960s that allowed Euroamerican participants, as a way of bridging "implacable social, racial, and temporal gulfs between Indians and modern non-Indians." Finally, some people view powwows through a financial lens, asserting that Native Americans created powwows "to fill the gaps in an uncertain local economy." In contemporary times, powwows also serve as a way in which Native Americans can compete in sport.

Categories of Events

Originally, there were only traditional powwows, which rewarded excellence with prestige. During the 1900s, however, another category of powwow emerged—the competition powwow. While traditional and competition powwows share many of the same features (e.g., drums, feathered regalia, dancing, and enthusiastic spectators and participants), competition powwows reward excellence with cash prizes. In competition powwows, male dancers compete in traditional, grass, and fancy dances, while female competitors take part in traditional, fancy shawl, and jingle dress dances. Specialty and exhibition dances are interspersed in the competition schedule to provide competitive dancers with much-needed rests. Some of these dances include the sneak-up dance, the blanket dance, and the hoop dance. The hoop dance, where the dancer uses twelve to twenty hoops to create different figures, is typically the crowd favorite.

Powwows, both traditional and, to a lesser degree, competitive, involve Aboriginal spiritual practices. Of particular importance is the ceremonial use of eagles and eagle feathers. The eagle, a warrior bird, "flies closest to the Great Spirit." In some Aboriginal cultures, the death of a family member was forecast when a feather fell from a dancer's regalia. Indeed, feathers add to the spiritual significance of powwows. The dance arena itself also takes on spiritual importance during a powwow. Once the dance area is blessed it becomes sacred ground.

Influence of Prize Money

For some, it is easy to see the logic behind competing to earn cash rather than prestige. Following the powwow circuit can be expensive, especially when one has to forgo other employment to do so. Earning cash prizes helps to ease the financial burden of powwow participation. Roberts found that a family on the powwow circuit can earn an average of four thousand dollars a month, which is significantly more than some families are able to earn by working on their reservation. Indeed some Native Americans feel that if Euroamericans can earn several million dollars a year by playing professional sports, there should not be anything wrong with their earning a few thousand dollars by participating in powwows. Certainly, professional powwow dancing can become a lucrative career path for a talented few.

Understandably, some powwow dancers and spectators feel that money has changed the nature of powwows. Commodification is increasingly prevalent in modern sport and has "drastically changed the nature of being an athlete." The commodification of powwow dancing has resulted in interesting occurrences, including cutthroat tactics among powwow competitors. Pringle tells of one instance where a competitor secretly sprinkled a strong drug on at least one powwow dancer's regalia, forcing him to withdraw from competition due to illness. Though instances of regalia tampering do occur, a more common complaint with powwows concerns standardization. Points are supposed to be awarded for stopping on a dime at the last drum beat, while points can be lost for failing to stop or for a feather falling from one's regalia. Nevertheless, Horse Capture finds that judging often takes place on an ad hoc basis.

Many contemporary powwows now have both traditional and competition portions, seeking to strike a balance between those who want to compete for cash prizes and those who do not. Though competition powwows exist on the periphery of what some may consider modern sport, the fact that they often draw hundreds, if not thousands of people, indicates that they should receive more attention from the world of sport than they currently do.

Audrey R. Giles

See also: All Indian Competitions; Heritage.

FURTHER READING

Deloria, P.H. *Playing Indian.* New Haven: Yale University Press, 1998.

Friesen, J.W. *Rediscovering the First Nations of Canada.* Calgary, ALTA: Detselig Enterprises, 1997.

Hall, A., T. Slack, G. Smith, and D. Whitson. *Sport in Canadian Society.* Toronto: McClelland and Stewart, 1991.

Horse Capture, G.P. *Powwow.* Cody, WY: Buffalo Bill Historical Center, 1989.

Moore, J.H. *The Political Economy of North American Indians.* Norman: University of Oklahoma Press, 1993.

Paraschak, V. "Variations in Race Relations: Sporting Events for Native Peoples in Canada." *Sociology of Sport Journal* 14 (1997): 1–21.

Pringle, H. "Ceremonial Circuit." *Equinox* (July/August 1987): 37–46.

Roberts, C. *Pow Wow Country.* Helena, MT: American and World Geographic, 1992.

White, J.C. *The Powwow Trail: Understanding and Enjoying the Native American Pow Wow.* Summertown, TN: Book Publishing, 1996.

Rod CURL

Born January 9, 1943
Golfer

Curl is a Wintu from Shasta County in northern California. His tribal name is Yoso, the Johnny-jump-up wildflower. Rod showed his hand-eye coordination playing baseball at Central Valley High School. He graduated in 1961 and worked in construction, the best available job for young men without college educations. At age nineteen he took up golf at the nine-hole Lake Redding course and then joined the Riverview Country Club.

Rod quickly got down to a two handicap, and by age twenty-five he could beat anybody in town. He finished eleventh in the National Amateur tournament in 1968 and won his PGA card the next year. He still had to earn one of fifteen qualifier spots before each event.

At five feet, five inches, and 160 pounds, the Wintu was one of the smallest players on the circuit. He was on top of his game in 1974, enjoying seven top-ten finishes and placing seventeenth on the money list. His lone tournament victory came at the prestigious Colonial National Invitational in Fort Worth, Texas. The PGA made it one of three "designated" events in which all 150 top players were required to compete. CBS also broadcast the tourney for the first time.

Tight, tree-lined Colonial becomes one of the world's toughest courses when a stiff wind blows. So it was in the final round, as Curl outdueled none other than Jack Nicklaus down the homestretch. The Golden Bear bogeyed the seventeenth hole, while Rod finished four under par at 276 to take the $50,000 first prize. Shasta County proclaimed "Rod Curl Day" on May 20, 1974.

Rod Curl, best known for his victory at the Colonial National Invitational, has more recently prospered on the Senior PGA Tour. *(AP/Wide World Photos)*

Curl's best showing after that was runner-up at the Kaiser International Open in 1975. Another of the day's top stars, Johnny Miller, won by three strokes. Rod remained a regular on tour through 1980, amassing lifetime earnings of more than $700,000. He now resides in southern Florida, teaching at Abacoa Golf Course. He still competes occasionally on the Senior Tour and goes back each year for the Colonial on a lifetime exemption. Sons Rod, Jr., and Jeff are also professional golfers.

Rory Costello

FURTHER READING

Cubbedge, Mark. "Teaching a New Tradition." *PGA TOUR.COM Online Magazine*, July 10, 2000.

Arthur Lee DANEY

Born July 9, 1904
Died March 11, 1988, Scottsdale, Arizona
Baseball player

Arthur Lee Daney (Choctaw) appeared as a pitcher for the Philadelphia Athletics during the 1928 season, marking the high point of an amateur, semiprofessional, and professional baseball career that spanned the 1920s and 1930s.

Born in 1904, Daney grew up in Talihina, Oklahoma, playing baseball and emulating the players on a local team. He attended the Jones Academy, a boys' school run by the Choctaw tribe in Hartshorne, Oklahoma. After earning his high school diploma, Daney pitched for Haskell Indian School in Lawrence, Kansas.

While a teenager, Daney spent summers playing semiprofessional baseball in and around Oklahoma. In 1927, he traveled to Kansas City and signed on with a Negro league team (possibly the league-dominant Kansas City Monarchs), then joined another club based in Concordia, Kansas. Daney had a good season, which culminated in a win at a tournament in Denver, Colorado, where he impressed Ira Thomas, a scout for the Philadelphia Athletics. After the game, Thomas offered Daney a contract to play for the Athletics the following year.

Daney reported to the Athletics' training camp in Fort Meyer, Florida, in the spring of 1928. He joined manager Connie Mack and a group of talented players in a team that was to be a dominant force in subsequent years. First basemen Jimmie Foxx and center fielder Al Simmons came up that year as rookies. The pitching staff included Lefty Grove and Jack Quinn (the last of the legal spitball pitchers). Mickey Cochrane was the catcher. Tris Speaker and Ty Cobb were also with the team, at the ends of their long careers. Cobb liked to talk to Daney about Indians and called him "Chief Coolem Off." The nickname preferred by Daney's teammates was "Chief Whitehorn," provided by Athletics coach Kid Gleason, who thought Whitehorn sounded "more Indian" than Daney.

Competition among the Athletics pitching staff proved to be too much for Daney, and he pitched only a handful of innings during the 1928 season. His first appearance—three innings of relief against the New York Yankees—was his most memorable. The first batter Daney faced was Babe Ruth, who hit a towering infield popup. Lou Gehrig came up next and stroked a long fly ball to center field. Daney retired the side by striking out Jumpin' Joe Dugan, then went on to pitch two more scoreless innings in a 9–6 Athletics loss.

For the 1929 season, Mack sent Daney to pitch for the Indianapolis Indians, the Athletics' top minor league club. Daney stayed with the Indians for three seasons, compiling a record of four wins and fourteen losses in thirty-seven appearances. He never again appeared in the majors but continued to pitch for a number of minor league and semiprofessional teams until 1939, before giving up baseball to return to Oklahoma and enter the ministry. In 1969, Daney moved to Scottsdale, Arizona, where he lived until his death on March 11, 1988, at the age of eighty-three.

Nicolas Rosenthal

FURTHER READING

Paine, Dru. "Arthur Lee Daney." In *Life and Times of the Original Choctaw Enrollees,* edited by Wesley Samuels and Charles Samuels. United States, 1997.
Wright, Marshall D. *The American Association: Year-by-Year Statistics for the Baseball Minor League, 1902–1952.* Jefferson, NC: McFarland, 1997.

Scott DANIELS

Born September 19, 1969, Mistawasis First Nation Reserve, Saskatchewan
Hockey player

Daniels (Cree) was born on the Mistawasis First Nation Reserve, a small all-Native community in north-central Saskatchewan that is home to some six hundred people and is forty miles from the closest urban center, Prince Albert. He rose through the ranks from shy community hockey player to National Hockey League (NHL) "tough guy" within a few short years.

Growing up in Mistwasis, Daniels childhood was consumed with hockey. Playing in local hockey leagues, his skills developed to the point where he had attracted the attention of professional scouts. At age seventeen Daniels began his professional career, with the Kamloops Blazers of the Western Hockey League (WHL). A grinding forward, early on Daniels established himself as an enforcer, first with the Blazers and then with the New-Westminister Bruins and the Regina Pats between 1986 and 1990. He developed into a power forward in the junior ranks, a player who could both score goals and take care of himself on the ice.

Soon after the Hartford Whalers made him the 136th overall pick in the 1989 draft, Daniels found himself with the Springfield Indians of the American Hockey League (AHL), where he would play for the next three years. Despite being an integral role player for the Indians in their consecutive Calder Cup winning season of 1990 and 1991, Daniels would have to wait until 1993 before he would play his first NHL game—the Whalers against the Boston Bruins—where he made an immediate impression by accumulating nineteen penalty minutes.

In addition to playing in Hartford, Daniel's career led him to brief stays with the Philadelphia Flyers and the New Jersey Devils. During his six-year NHL career Daniels scored eight goals and twenty points while registering 667 penalty minutes. In one game alone, he accumulated twenty-nine minutes in penalties. Like many Native players in professional hockey, Daniels accepted his enforcer role. As a member of the "Destruction Line" while in Hartford with line mates Mark Janssens and Kelly Chase, Daniels became a fighter who policed his team's star players, making sure they were not intimidated. In one memorable 1996 fight with Rob Ray, Daniels, as a member of the Whalers, found himself in the Sabres' bench during the brawl and simply came out swinging at any player he did not recognize as a team mate.

Daniels retired from the professional ranks in 1999 after being unable to secure a position with an NHL club and today lives with his wife and children in Agawam, Massachusetts. Family is an important aspect of Daniels's life, which was evidenced by his missing an April 18, 1995, game to be with his wife for the birth of their daughter. Each summer he directs the Scott Daniels Hockey School in Springfield, Massachusetts.

Yale D. Belanger

FURTHER READING

Mistawasis First Nation (www.sicc.sk.ca/bands/bmista.html).
Hockeydb.com, "Scott Daniels" (www.hockeydb.com/ihdb/stats/pdisplay.php3?pid 251).

Ron DELORME

Born September 3, 1955
Hockey player

Delorme (Cree) became involved in hockey at a very early age, an interest that later

During his time, Ron Delorme was known as the hardest working wing in hockey. *(AP/Wide World Photos)*

translated into a professional career that has now spanned three decades.

Delorme began his professional playing career at the age of eighteen, with the Swift Current Broncos of the Western Canadian Hockey League in 1973–1974. During this period, the emerging World Hockey Association and the National Hockey League were competing for players, and Delorme found himself drafted by both the Kansas City Scouts of the National Hockey League and the Denver Spurs/Ottawa Civics of the World Hockey Association. After playing twenty-two games in the fledgling World Hockey Association he was sent down to the minors for one more year.

The next season Delorme returned to professional ranks with the Colorado Rockies of the National Hockey League, with whom he would remain until joining the Vancouver Canucks in 1981. Considered at the time to be one of the hardest-working right wingers in hockey, Delorme finished his nine-year National Hockey League career with the Canucks in 1985 after a knee injury, tallying eighty-three goals and 166 points in 524 games with 667 penalty minutes.

A grinding forward, Delorme played a key role in the Canuck's Stanley Cup run of 1982 and is probably best remembered for his exhilarating fight with Chicago Black Hawks' Grant Mulvey in the Campbell Conference Final. Following his retirement, Delorme continued with the Canucks in a variety of positions, working primarily as a scout for the next fifteen years. In 2000, he was promoted to chief amateur scout, coordinating the Canuck's amateur scouting staff and assembling the team's draft selection list.

In addition to his scouting duties, Delorme cofounded (with Kevin Tootoosis) the Aboriginal Role Models Hockey School in the late 1990s. Having wondered why there were so few professional Aboriginal hockey players, the two men decided to open the school to aid young players in their transition to professional hockey and to provide a forum in which to demonstrate their skills to professional scouts. With the sponsorship of the National Hockey League Players Association, which provides equipment and pays for ice time, the Hockey School continues to provide leadership and help ease the anxieties of aspiring Aboriginal hockey players.

"Because of the fact that Native kids' upbringing is a little bit different on the reserves, there's nobody to tell them to watch for the green light, the red light, the yellow light as far as crossing the road, and that's just a little example," Delorme stated in an interview with the *Saskatchewan Sage* in 2000. He added, "Our upbringing is a lot different than it is in the city. They have to understand [city] culture."

Yale D. Belanger

FURTHER READING

Roden, Marjorie. "Role Models Hockey School Keeps Hockey Growing." *Saskatchewan Sage* (www.ammsa.com/sage/AUG2000.html), August 21, 2000.

NHL.com Network. "Ron Delorme" (www.canucks.com/subpage.asp?sectionID 93#title).

William "Lone Star" DIETZ

Born 1885, Rosebud Reservation, South Dakota
Died 1964
Football coach

William Dietz was a true renaissance man. He drew illustrations for magazines and books; he taught art; he acted in movies; he played football; and he coached football. Maybe more so than other prominent Native American athletes, Dietz is known on and off the football field.

Dietz was born in 1885 on the Rosebud Reservation in South Dakota. He was later known as William Dietz, but his given name in Sioux was Wicarphi Ismala (Lone Star). Dietz went to grammar school in Rice Lake, Wisconsin, and developed an interest in art. Dietz attended Carlisle Indian School in 1907 and married Angel DeCora (Winnebago) in 1908. She taught art at Carlisle and was fifteen years Dietz' senior. At Carlisle, Dietz blossomed into a popular artist. His drawings appeared on the cover of *The Arrow* and *The Red Man*, student publications at Carlisle. The content of his drawings ranged from the assimilation of Indians into American society to Native American life before European contact.

Dietz also showed promise on the gridiron. He began playing football in 1909; head coach Glenn "Pop" Warner inserted him into the starting lineup at defensive tackle. After Dietz finished playing for the Indians, Warner hired Dietz as an assistant coach, for $500 per year. While Dietz served as Warner's understudy, he taught design at the School of Applied Art in Philadelphia.

Beginning in 1915, Dietz embarked upon a peripatetic coaching career. Between 1915 and 1918, Dietz coached Washington State College in Pullman, Washington. During his inaugural campaign, Dietz led the team to an undefeated record and a 14–0 victory over Brown University in the first East-West game, later renamed the Rose Bowl. During Dietz's three seasons, Washington State lost only two games. While at Washington State, Dietz divorced DeCora (she returned to the East Coast). In 1921, Dietz coached Purdue University to a 3–5 record. Between 1922 and 1926, Dietz coached Louisiana Polytechnic Institution and the University of Wyoming.

In the mid-1920s, Dietz came under Warner's wing again. He drew the illustrations for Warner's book *Football for Coaches and Players.* Dietz also won a job coaching the freshman team at Stanford University. While in California, Dietz also starred in a few movies. Beginning in 1929, Dietz began a three-year tenure as the head coach at Haskell Indian School. Between 1932 and 1935, Dietz coached the Boston Braves of the National Football League. Under Dietz, the Braves changed their name to the Redskins and later moved to Washington. After Redskins' management replaced their Native American coach, Dietz joined Warner's staff at Temple University. He finished his twenty-seven-year coaching career with five years at Albright College in Pennsylvania.

Dietz's life after football was not as successful as it had been while he was engaged in football. During World War II, Dietz worked for an advertising firm in New York. Afterward, he made numerous financial mistakes and had to scramble to

William "Lone Star" Dietz (middle) poses with his coaching staff at Haskell Institute, John Levi (left) and Egbert Ward (right). *(AP/Wide World Photos)*

make money in his twilight years. Dietz died in 1964 from cancer. In 1997, Dietz was elected to the Pennsylvania Sports Hall of Fame.

William J. Bauer, Jr.

FURTHER READING

Ewers, John. "Five Strings to His Bow: The Remarkable Career of William (Lone Star) Dietz." *Montana* 27 (January 1977): 2–13.

Newcombe, Jack. *The Best of the Athletic Boys: The White Man's Impact on Jim Thorpe.* New York: Doubleday, 1977.

DOGSLED RACING

Dogsled racing utilizes a harnessed team of dogs, one of which is trained as a leader. The team is managed by a driver (called a "musher") and pulls a sled along a trail or track, usually covered with snow or ice. The competition generally involves speed, distance, or a combination of the two. In the northern parts of North America, beginning perhaps as early as four thousand years ago, dogsleds enhanced travel during the nine or ten months of the year when snow covered the ground and ice covered the seas. Dogsleds were used to carry household goods from one campsite to another and to carry meat from the kill site to camp.

History of Racing

The first organized dogsled race was probably the All-Alaska Sweepstakes, held in April 1908 in Nome. Sponsored by the Nome Kennel Club, it was run from Nome to Candle (204 miles to the northeast) and back. Many of the rules developed for this race are still observed. It became an annual event and lasted through 1917. Canada's organized races began even earlier (1857) with a four-hundred-mile freight race (each team pulled three hundred pounds) from Fort Garry, Manitoba, to Crow Wing, Minnesota.

Early U.S. races outside Alaska included the American Dog Derby in Ashton, Idaho, begun in 1917 and run almost every year until 1948. The New England Sled Dog Club, perhaps the oldest club in continuous existence, was organized in 1924 and sponsored two races the next year. Dog breeders began producing smaller, faster dogs, and races generally changed from long-distance marathons like the All-Alaska to shorter courses run over two or three days.

The 1928 Olympics, held in St. Moritz, Switzerland, included a three-day, 123-mile race. Dogsled races were also held as a demonstration sport during the 1932 Winter Olympics at Lake Placid, New York. The sport experienced a hiatus during World War II; in fact, the interval between the end of the All-Alaska Sweepstakes (1917) and the beginning of the Fur Rendezvous Championship (1946) saw few serious races that drew significant publicity. The International Sled Dog Racing Association was founded in 1966. It assists the sport by overseeing rules and management procedures and by maintaining the race sanctioning system, animal welfare standards, and race software.

By the early 1960s, snowmobiles had largely supplanted dogsleds among Natives and non-Natives. Joe Redington, in an effort to revive interest in dog mushing and preserve the Iditarod Trail (completed in 1910 to remove gold from the area), developed and promoted the Iditarod Trail International Sled Dog Race, which has become the sport's most widely known event. Alaska's state legislature proclaimed dog mushing the official state sport in 1972. The first full-length Iditarod began on March 3, 1973, with thirty-four teams and a large purse, fifty thousand dollars. The 1,049-mile annual race from Anchorage to Nome has had participants ranging in age from eighteen to eighty-eight and from thirteen other countries. In 2003, sixty-four mushers paid an entry fee of $1,850 to compete. The total purse was six hundred thousand dollars.

Native Participation

Involvement of Native peoples in dogsled racing continues to be strong. Native participation is higher in Alaska than in the lower forty-eight states. In Canada, Native peoples are heavily involved, especially in the Northwest Territories. Native mushers in the 2003 Iditarod included Mike Williams, Palmer Sagoonick, and Ramy Brooks. Several Native teams competed in the 2003 Open North America Championships in Fairbanks, Alaska. Some of the most revered Native mushers have been George Attla, Herbie Nayokpuk, and Emmitt Peters.

Importantly, some believe that the indigenous sport has been taken over by non-Natives. In Alaska, few Natives travel to larger towns to compete. Many, however, participate in local races, particularly during spring carnivals. By one estimate only about fifty indigenous teams are actively racing at present. Similarly in Canada, Native participation in racing as a leisure activity and cultural event is much higher than in formal races with large purses. Small towns and villages hold races throughout the winter, promoting them as part of their heritage and culture. Fred Hems, vice president of the Ma-Mow-We-Tak Sled Dog Association (which represents western Canada and the Northwest Territories), reports that at least 25 percent of its 120 members are Natives. In north-central Manitoba and Saskatchewan, at least 50 percent of the dogsled racing community at large is Native (the percentage climbs to 95 in the northern parts of these provinces). In the Northwest Territories, Native racers constitute 50 percent in the southern region and higher farther north.

Glenn Ellen Starr Stilling

FURTHER READING

Anderson, Dennis. "Life at the Top of the World Is Isolated, Difficult, Steeped in Tradition." *Star Tribune,* February 16, 2003, B12.

Cellura, Dominique. *Travelers of the Cold: Sled Dogs of the Far North.* Anchorage, AK: Northwest Books, 1990.

Coppinger, Lorna. *The World of Sled Dogs: from Siberia to Sport Racing.* New York: Howell Book House for the International Sled Dog Racing Association, 1977.

Freedman, Lew. *Father of the Iditarod: The Joe Redington Story.* Fairbanks, AK: Epicenter, 1999.

Sprott, Julie E. "Christmas, Basketball, and Sled Dog Races." *Arctic Anthropology* 34, no. 1 (1997): 68–85.

DOUBLE BALL

Double ball, which is similar to lacrosse, was also called the maiden's ball play, twin ball, and the women's ball game. Several other names were also used among various tribes in different regions of the United States. Double ball was considered a woman's game, but men played it in a few northern California tribes.

In double ball, a game could be played with as few as six on each team, but teams of up to 100 were not unusual. The double ball consisted of two balls or sticks attached by a string or thong. The object of the game was to advance toward the opponent's goal by throwing and catching the double ball with a stick.

Normally three to four feet long, the double-ball stick was commonly made of a thin, flexible willow branch. On one end it was curved, which allowed players to control the ball with some accuracy. For beginners, the stick had a notch near the end to make it easier to control the ball. These sticks were often decorated with carvings, paintings, and symbols. Among some tribes, these markings were burned into the wood.

The double balls were constructed to be caught on one end of the stick, with the two balls falling on different sides of the stick. Two basic types of buckskin balls were used, although the balls varied in shape and construction among different tribes. One version was two separate balls made of buckskin and attached by a thick string or thong. Another version was a dumbbell-shaped buckskin ball with a single piece of leather, heavy at each end and thin in the center. These were filled with sand or other material to weight down the ends. Another kind of double ball was made of two sticks of wood attached together by a heavy string. These sticks were six to ten inches long and were held together by cords six inches to one foot in length. As in other ball games, the balls were brightly decorated.

Most playing areas were 300 to 400 yards long, but some were reported to be a mile or more in length. At each end of the field, various objects were used as goal markers, including trees, poles, and piles of dirt. Placed twelve to fifteen feet apart, these markers were also called base, home, or goal. Among the Cree Indians, double ball was played only by young women, as great powers of endurance were needed. To begin the game, the players gathered in a circle, and the double ball was thrown from the stick of one of the leaders. The game continued until one side passed the ball through the opponent's goal. Like many other Indian activities, double ball was played for stakes of some kind. Descriptions of double ball indicate that women took sport participation seriously and played with a high level of physical skill and enthusiasm.

Susan Keith

See also: Gender Relations; Lacrosse; Women.

FURTHER READING

Culin, Stewart. *Games of the North American Indians.* Washington, DC: Government Printing Office, 1907.

Oxendine, Joseph B. *American Indian Sports Heritage.* Champaign, IL: Human Kinetics Books, 1988.

Vallie Ennis EAVES

Born September 6, 1911
Died April 19, 1960
Baseball player

Eaves played high school baseball in Rosedale, Oklahoma. In 1927, this right-handed pitcher became involved in a "schoolboy scuffle" that resulted in a shinbone infection so severe that he remained hospitalized for nine months. A doctor was quoted as saying, "You'll never pitch again. It's improbable that you'll ever walk." Even though his right leg continued to trouble him and he required crutches for four years, Eaves returned to baseball, pitching for semipro teams in 1933 and 1934.

In 1935, Eaves married Lorraine Martin and pitched his first professional game for the Philadelphia Athletics, notching a win against the Chicago White Sox. After losing a couple of games, however, he was sent home. Eaves noted that coach Connie Mack did not want to take a chance on his leg, which had developed osteomyelitis. After that, Eaves wore a leg brace.

Returning to semiprofessional baseball, the Texarkana team offered this six-foot, two-inch pitcher a contract in 1938, the year that Vallie "Chief" Eaves earned 209 strikeouts and posted twenty-one wins. "Speed," writes one reporter, "burning, blazing speed, has been Eaves' main reliance." Years on crutches, though, caused permanent stiffness in his wrist, eliminating the curve ball from his repertoire. "He was forced to rely on speed alone, plus a knockle [*sic*] ball he developed as a substitute for the curve." Another reporter wrote that Eaves had the "speed to make a ball look the size of BB shot when it bisects the plate."

In 1939, Eaves pitched for Shreveport before being picked up by the Chicago White Sox, for whom he played through 1940. He pitched for the Chicago Cubs in 1942 and continued to pitch for semiprofessional teams through 1957. That year, he played on the Hobbs team with his son, Jerry.

Early in his career, Eaves earned the nickname of "baseball's bad boy." In 1942 he was suspended by the Cubs, accused of intoxication at the ballpark. In 1945, manager Pepper Martin slugged Eaves in a hotel lobby, accusing him of losing games because he was hung over. In 1959, when Eaves was battling cancer, he was quoted as saying, "I got into a lot of trouble because of drinking, but I wasn't any worse than a lot of others. I just got caught, that's all."

At the age of forty-eight, Eaves died of lung cancer. He was survived by his wife, three sons, and a daughter.

Kelly Boyer Sargent

FURTHER READING

The Ballplayers Mini Biography (www.baseball-almanac.com/players/player.php?p avesva01).
Baseball-Reference.com (www.baseball-reference.com/e/eavesva01.shtml).

Mike EDWARDS

Born December 14, 1961, Oklahoma City, Oklahoma
Bowler

Edwards is a popular and accomplished bowler. Edwards is the only Native American bowler in the Professional Bowlers Association (PBA). He is also the only bowler enshrined in the American Indian Athletic Hall of Fame.

Edwards turned professional in 1982, joining the PBA Tour. For over a decade, he endured numerous heart-breaking losses and near victories. In fact, for a time he had a reputation as a skilled competitor who always came up just a little bit short. For instance, in 1989, he was the runner-up at the ABC Bud Light Masters. Five years later, in 1994, he won the IOF Foresters Open. Although he only has one title to his name, two other factors attest to Edwards' excellence as a bowler: money and television appearances.

After 20 years on the PBA tour, his career earnings approach the $1,000,000 mark. For 15 years, he has ranked among the top 30 money winners and his lifetime earnings of $824,221 (as of 2003) places him in the top 40 all-time money winners.

Edwards has appeared regularly on national broadcasts of the PBA. During his career, he has appeared more than 25 times and he has the distinction of appearing on five different networks (ABC Sports, ESPN, Prime Sports Network, Sports Network, and USA Network).

Edwards has bowled 20 sanctioned 300 games. With an 869 average, he has the record for the highest score in a three game series in Oklahoma.

Edwards is proud of his heritage, working actively with a number of Native American organizations, and serving on the ABC Minority Concerns Committee. He resides in Oklahoma City, where he is a teaching pro at Heritage Lanes.

C. Richard King

FURTHER READING

PBA (www.pba.com).

Fait ELKINS

Born 1905, Anadarko, Oklahoma
Died 1966, Philadelphia, Pennsylvania
Football player, basketball player, decathlete

Fait Elkins was one of the most accomplished track and field stars of the 1920s, while also performing brilliantly in football and basketball; he was hailed by many sports writers in his day as the best all-around Native American athlete after Jim Thorpe.

Elkins was born in Anadarko, Oklahoma. His father, Stephen Elkins, was a full-blooded Caddo Indian, and his mother, Fannie Mays, was white. When he was in his early teens, from 1922 to 1924, he attended Haskell Institute, an Indian school in Lawrence, Kansas. Haskell was basically a preparatory school, but its varsity teams often competed against colleges, and in 1922 and 1923 Elkins, despite his youth, showed considerable ability running and kicking the ball on the school's football team. In one game he scored five touchdowns.

In early 1924, Elkins was lured to Southeastern State Teachers College in Durant, Oklahoma. Although not enrolled at the school, under the name of "The Chief," he started on the basketball team and led his team to several victories in track and field. While at the Oklahoma school, he also competed on the Haskell track and field team. In the fall, Elkins—now officially enrolled at Southeastern—played football and helped the Durant school win its league title. From December 1924 through the fall of 1925, Elkins starred in track and football at Dallas University.

In the fall of 1926, the itinerant Elkins entered the University of Nebraska (in Lincoln) as a freshman and by the spring was starring on the Cornhuskers track team. In July 1927, the Amateur Athletic Union

(AAU) national track and field meet was held in Lincoln, Nebraska, and Elkins won the decathlon title with a score of 7574.42, close to a U.S. record. Elkins's score was remarkable because the AAU held the decathlon on one (brutally hot) day, instead of the usual two. Elkins' athletic career at Nebraska ended the fall of 1927, when the conference found him ineligible on the grounds that he had already played five years of college football.

In 1928, Elkins performed for the New York Athletic Club in a practice meet in June, posting a decathlon score of 7.802.17 that would have been a new American record under official conditions. The newspapers were likening him to Jim Thorpe and promoting him as America's best hope to win the decathlon in the upcoming Olympics. However, at the Olympic trials in July, Elkins pulled up lame in the 100 meters and could not finish. He was forced to sit out the Olympic Games.

Elkins entered professional football in October 1928, joining the Frankfort Yellow Jackets as a running back. With Elkins all-around abilities (he kicked off, punted, and ran), the Yellow Jackets garnered an 11–3–2 record, good for second place in the league. His most spectacular play was a kickoff return of ninety-eight yards in a 19–0 victory over the Chicago Cardinals. The next year he started the season with the Chicago Cardinals, but ended the season with the Yellow Jackets.

In 1931, Elkins led an all-Indian barnstorming team, the Hominy Indians. In 1933, he played his final season in the NFL, for the Cincinnati Reds. Elkins died in Philadelphia.

Robert Pruter

FURTHER READING

Gilbert, Bill. "The Twists of Fate: An Athlete of Rare Talent, Fait Elkins Seemed Destined for Immortality but Landed In Obscurity." *Sports Illustrated Classic* (Fall 1995): 77–81.

Zarnowski, Frank. "Fait Elkins." In *Olympic Glory Denied: And a Final Opportunity for Glory Restored.* Glendale, CA: Griffin, 1996.

EQUESTRIAN COMPETITIONS

American Indian equestrian competitions consist primarily of horse racing and rodeo events. Indians began racing horses not long after they acquired them from Spanish explorers and settlers in the late seventeenth to mid-eighteenth centuries. Many Indians considered this acquisition to be the reintroduction of the horse; Navajo and Apache myths tell of a rider-hero who borrows Father Sun's fastest horse in order to beat his opponent in a race around the edge of the world. For Indians, horses were gifts from the Creator, sacred beings possessing supernatural powers, or bridges by which Indians could access supernatural powers. Indians formed both social and religious relationships with horses; horses were revered and honored. Horses continue to be venerated in many contemporary tribal cultures.

In the eighteenth and nineteenth centuries an exceptional Plains racehorse was prized more for its stamina and endurance than for its speed, due to the horse's use in buffalo hunting and warfare. In Blackfoot and other Plains and Plateau Indian horse cultures, highly trained horses could be turned simply by pressure from a rider's knee or by the rider's shifting his weight from one side to the other. Good racehorses were typically small and compact, often no more than fourteen hands in height. Most races were match races, with one horse and rider competing against another horse and rider. Races were run against fellow tribes

Following the introduction of the horse, equestrian competitions, like this one in Fruitland, New Mexico, became important in many Native American communities. *(Denver Public Library/Western History Department)*

people or against members of other tribes. The most common races were between two societies of the same tribe; generally, men of the same society did not compete against each other. The typical racecourse was a straightaway up to three or four miles in length; some races were run on mountainous courses. Jockeys were generally young men, but women also raced horses. Racehorses were usually males, considered to be in their prime from the ages of four to nine years. These horses were not used for packing or general riding but were employed after their racing careers as buffalo runners. In Blackfoot culture a winning racehorse was a person's most valuable possession. Stakes in horse races could be very high and included possessions, money, and prestige.

In the nineteenth century, Indians also raced their horses against Euro-American explorers, trappers, and traders, and U.S. soldiers. During the spring of 1806, Nez Perce Indians demonstrated their superb horses and horsemanship in races against members of the Lewis and Clark Expedition. After the near extermination of the buffalo and the Indians' confinement to reservations in the later nineteenth century, horse racing continued to a limited extent on reservations. Government agents often allowed Indians to practice ceremonial activities on American holidays, such as the Fourth of July, with horse racing a major activity. Many tribal fairs began in the twentieth century; due to the attraction of horse racing, races became an integral part of fairs, rodeos, and powwows. The first Crow Fair, in Montana in 1904, was modeled on midwestern agricultural fairs. Instead of inspiring pride in agricultural achievements, however, as the reservation's Indian agent had hoped, the fair revived Indian rituals, including horse races, with a track changed from a straightaway to a circle.

In the early to mid-twentieth century, as Indians became ranchers and cowboys, their equestrian expertise proved extremely valuable. Rodeo, the sport of ranching culture, provided numerous opportunities for Indians to demonstrate their equestrian skill and to reaffirm their identities as Indians. Virtually all competitive events in rodeo, except bull riding, rely heavily on a rider's equestrian skills. The saddle bronc-riding event emulates the breaking or taming of a wild horse and requires a keen sense of a horse's bucking rhythm. Bareback bronc-riding has no direct corollary to ranch work and was developed specifically for competition. It requires an eight-second ride on a bucking horse while holding a leather grip with one hand. The rider's free hand can touch neither himself nor the horse. Calf roping depends upon a successful working relationship between rider and horse. A rider on horseback must rope a running calf that has been given a head start, secure the rope to his saddle, dismount, run down the calf, throw him to his side, and tie three of the calf's legs together with a string he had held in his mouth. If the rider is to accomplish this, the horse must keep the rope holding the calf taut by moving backward. Other equestrian rodeo events include steer wrestling, steer roping, and team roping. Racing events include barrel racing, wild horse and chariot (or chuck-wagon) racing.

Although a few Indians, such as Jackson Sundown and Tom Three Persons, have been highly successful in mainstream rodeo, a tradition of all-Indian rodeos began developing in the mid-twentieth century due to the discrimination and exploitation Indians often faced when participating in non-Indian rodeos. Indian rodeos highlight the honor accorded animals by Indians and the close relationship that exists between Indians and animals, particularly horses. In Indian rodeo, animals and humans work together to achieve athletic excellence. The first Indian rodeo association, the All Indian Rodeo Cowboys Association, was founded in 1957 and is now called the All Indian Professional Rodeo Cowboys Association. The Indian National Finals Rodeo is held each summer, while many smaller rodeos are held throughout the West and Southwest, including the Crow Fair Rodeo in Montana, the Navajo Fair and Rodeo in Arizona, and the Inter-Tribal Ceremonial Rodeo in New Mexico. The Indian Junior Rodeo Association reflects rodeo's growing popularity among children and young people. The sport of rodeo continues to increase in importance for Indians, due to the opportunity it provides for expressing and preserving Indian identity and for demonstrating native peoples' athletic excellence. Although horse racing survives to a lesser degree in places like the Navajo Nation, rodeo has become the most popular and vibrant of Indian equestrian competitions.

Janis A. Johnson

See also: All Indian Professional Rodeo Cowboys Association; Rodeo; Wild West Shows.

FURTHER READING

Baillargeon, Morgan, and Leslie Tepper. *Legends of Our Times: Native Cowboy Life.* Seattle: University of Washington Press, 1998.

Clark, Laverne Harrell. *They Sang for Horses: The Impact of the Horse on Navajo and Apache Folklore.* Albuquerque: University of Arizona Press, 1966.

Dobie, J. Frank. *The Mustangs.* Boston: Little, Brown, 1934.

Ewers, John C. *The Horse in Blackfoot Indian Culture.* Washington, DC: Smithsonian Institution Bureau of American Ethnology, 1954.

Iverson, Peter. *When Indians Became Cowboys: Native Peoples and Cattle Ranching in the American West.* Norman: University of Oklahoma Press, 1994.

Iverson, Peter, and Linda MacCannell. *Riders of the West: Portraits from Indian Rodeo.* Seattle: University of Washington Press, 1999.

Josephy, Alvin M., Jr. *The Nez Perce Indians and the Opening of the Northwest.* Boston: Houghton Mifflin, 1965.

Lawrence, Elizabeth Atwood. *Rodeo: An Anthropologist Looks at the Wild and the Tame.* Knoxville: University of Tennessee Press, 1982.

White, Richard. "Animals and Enterprise." In *The Oxford History of the American West,* edited by Clyde A. Milner. Vol. 2. Oxford: Oxford University Press, 1994.

Darrell EVANS

Born May 26, 1947
Baseball player

Evans (Yavapai) was a record-setting major league baseball home run hitter. The first player to hit forty homers in both the National and American Leagues, Evans is also the oldest player to win a home run title, hitting forty homers for the Detroit Tigers at age thirty-eight in 1985.

Initially signed to a minor league contract by the Kansas City Royals, Evans in 1967 played his first season of professional baseball for minor league teams in the Carolina, Florida State, and Gulf Coast Leagues. In 1968, he played for Birmingham, Alabama, in the Southern League. In 1969, he made his major league debut with the Atlanta Braves, also playing that year for the Braves' minor league teams at Shreveport, Louisiana, in the Texas League and Richmond, Virginia, in the International League. In 1970 and 1971, he played for Atlanta and Richmond.

In 1971, Evans became the starting third baseman for Atlanta and hit twelve home runs. In 1973, the Braves had three record-setting sluggers with at least forty homers—Dave Johnson (forty-three), Darrell Evans (forty-one), and Hank Aaron (forty).

In June 1976, Evans was traded in a six-player deal to the San Francisco Giants of the National League. Although he was productive and hit 132 home runs in seven seasons for the Giants, Candlestick Park was not suited for his power. In 1983, the Giants switched Evans to first base, where he played 113 games. On June 15, 1983, he hit three homers in a game. After the season ended, he signed as a free agent with the Detroit Tigers of the American League.

Detroit got immediate dividends when Evans hit a three-run homer on opening day of the 1984 season. The Tigers coasted to the American League pennant and World Series championship, defeating the National League's San Diego Padres in five games. Overall, Evans had a poor-hitting year in 1984, including the World Series. In 1985, he bounced back with forty home runs to lead the league, beginning the most productive power stretch of his career. As a full-time designated hitter in 1988, he managed twenty-two homers but hit only .208 and was traded by Detroit back to Atlanta during the off-season. In 1989, he completed his career with Atlanta, hitting eleven home runs for a career total of 414, a figure that at his retirement was twenty-second all-time.

Royse Parr

FURTHER READING

Thorn, John, Pete Palmer, Michael Gersham, and David Pietrusza. *Total Baseball.* New York: Viking, 1997.

Albert Andrew "Exie" EXENDINE

Born January 7, 1884
Died January 4, 1973
Football player and coach

Albert Andrew Exendine, college football player, track athlete, and coach, was considered to be one of the top defensive ends of his time. Later, he compiled a long football coaching career at several schools.

Exendine, a Delaware Indian, attended the Mautame Presbyterian Indian Mission school before arriving at the Carlisle Indian School in 1899 at the age of fifteen. Never having played football previously, Exendine joined the Carlisle varsity team in 1902 and initially played tackle. By 1904 he had earned a starting tackle position, but he was shifted to end in 1905 in order to take advantage of his speed and athleticism. Exendine was also an outstanding track and field athlete; he set school records in several events before the arrival of Jim Thorpe at Carlisle.

For the 1906 season Exendine was elected team captain and became recognized as one of the best defensive ends in college football. His 1906 season was highlighted by an eighty-yard touchdown run after recovering a fumble against Pennsylvania, and in the postseason he was awarded a spot as third-team end on Walter Camp's All-American squad.

Exendine compiled an even more brilliant season in 1907, his sixth and final year playing varsity football for Carlisle. One of the top performances of his college career came that season against the University of Chicago. That day Exendine hauled in a fifty-yard touchdown pass from Pete Hauser; even more significantly, he played a sensational game on defense, repeatedly tackling, and shutting down the offensive threat of, Chicago's All-American halfback Walter Steffen. For the 1907 season Exendine received first-team All-American end berths from selectors Caspar Whitney and the *St. Louis Star-Chronicle,* along with a second-team spot from Walter Camp.

After graduating, Exendine served as an assistant coach at Carlisle in 1908 under Glenn ("Pop") Warner. He then began his career as a head football coach at Otterbein College in Ohio, where he served from 1909 through 1911. After receiving his law degree from Dickinson College in 1912, Exendine returned to Carlisle as an assistant coach in 1913 before assuming the head football coach position at Georgetown University in Washington, D.C., in 1914. He remained at Georgetown through the 1922 season; then followed head coaching jobs at Washington State (1923–1925), Occidental College (1926–1927), and Northeastern Oklahoma State (1928).

Exendine served as an assistant coach at Oklahoma A & M from 1929 through 1933, during which time he also was the head baseball coach for the 1932 and 1933 seasons—compiling an overall diamond record of nineteen wins and thirteen defeats. He concluded his collegiate football coaching career at Oklahoma A & M during the seasons of 1934 and 1935. In his twenty seasons as a gridiron coach he produced an overall career record of 93–60–15.

For years Exendine had worked during the football off-seasons as an attorney in Oklahoma, and after his retirement from college coaching in 1936 he continued his law practice, while also beginning a career as an organizational field agent for the Bureau of Indian Affairs in Oklahoma. After his retirement from the bureau in 1951, Exendine and his wife, Grace (the couple had one son), settled in Tulsa, Oklahoma.

Exendine was the recipient of numerous awards throughout his life, highlighted by his induction into the College Football Hall of Fame in 1970 and the American Indian Athletic Hall of Fame in 1972. He is also a member of the Oklahoma Athletic Hall of Fame.

Raymond Schmidt

FURTHER READING

Johnson, John L. "Albert Andrew Exendine: Carlisle Football Coach." *Chronicles of Oklahoma* 43 (Autumn 1965): 319–31.

McCallum, John, and Charles H. Pearson. *College Football U.S.A. 1869–1972.* New York: Hall of Fame, 1972.

Steckbeck, John S. *Fabulous Redmen: The Carlisle Indians and Their Famous Football Teams.* Harrisburg, PA: J. Horace McFarland, 1951.

FANS AND SPECTATORS

Fans and spectators are an important part of any sporting event, from recreational to amateur to professional. Historically, sports provided students at boarding schools occasions to get together, mingle, have fun and socialize, and today they remain a source of pleasure, fun, and inspiration for fans throughout Native America. For both Native American boarding school students and for Native Americans as a whole, sporting events have called for a celebration of intertribal cooperation and identity, occasionally on a grand scale. The appeal of sports for many Native American spectators lies in both the historical memories and the reaffirmation of identity.

It is nearly impossible to discuss fans and spectators of these games and sports without discussing gambling, as it is an integral part of Native North American games. In fact, formal competitions were often characterized by gambling and betting. This often reached such a large scale that every spectator seemed obligated to bet on the outcome of the game. Athletes might participate in the sport in order to compete and receive prizes, but it could be said that spectators participated by gambling. Fans might bet a knife, food, jewelry, or anything else of value, including services, which were placed in large scaffolds constructed at each end of the field. Once the game ended, the winning team collected the contents of both scaffolds and headed home.

White spectators unfamiliar with Indian games and sporting events would shudder at the amount of violence in them. Furthermore, as one recent historian notes, "the game, innocent in itself, was generally attended with much bad behavior, drunkenness and licentiousness." One game in 1825, near Springfield, a mission in northern Georgia, had about 3,000 in attendance and the bets and wagers made during this game topped $3,500.

The Eastern Cherokee ritualized the spectators' wagering. Before bets were placed at midfield, each team would line up in their goals, with the fans behind them. The two teams would march to the center of the field, carrying the items to be wagered, with the fans marching behind them. At centerfield, fans and players would intermingle, agreeing on the bets to be placed. At the end of the game, spectators would, for the most part, dispassionately claim their winnings and await a rematch, when they could recoup their losses. As wins and loses balanced over time, there was generally no urge for retaliation before the next rematch.

Dispassion was not always the rule among spectators. The Oklahoma Choctaw deliberately did not bring weapons to matches as a precaution against violence. Furthermore, it should be noted that bribery was a concern when the stakes were too high, and this problem carried across nearly all tribes. The Creek and Choctaw tribes occasionally suffered such tension that games were canceled and issues settled violently—once with more than 500 people fighting, some to the death.

Large-scale wagering among Native American fans was clearly a prominent pastime; however, enticing the white man to place bets was enjoyable as well. At games, for example, women, some carrying babies, would cluster in groups, allowing Englishmen to intermingle and become more involved in the game. Occasionally, the women would place bets with one another, in the hopes that this would encourage more Englishmen to do the same. Betting and gambling on Native North American Indian sports was a big draw for white men, especially soldiers. However, this was not always well accepted. Government officials were highly concerned about this gambling and considered it to be a hindrance to educational and religious progress. In 1898, the state of Mississippi outlawed gambling at all Indian lacrosse games, cockfights, and duels; thereafter, as hoped, the church became the new focal point of tribal social life. This is just one example of action taken to socialize American Indians into the mainstream white society.

Fandom was a chance for the Native American to express his or her identity and to enjoy a sport at another level—for example, by placing bets. Race, religion, culture, and family are all linked with sports, and fans were well aware of this. Sports and fandom not only provided Native Americans with a chance to express their own identity but gave them the chance to become more socialized with the white mainstream society. Fandom offered a chance to beat the white man at his own game, it offered an education, and perhaps most importantly, it offered fun and sociality.

Beth Pamela Jacobson

See also: Gambling.

FURTHER READING

Blanchard, Kendall. *The Anthropology of Sport.* Westport, CT: Bergin and Garvey, 1995.

Bloom, John. *To Show What an Indian Can Do: Sports at Native American Boarding Schools.* Minneapolis: University of Minnesota Press, 2000.

Vennun, Thomas, Jr. *American Indian Lacrosse: Little Brothers of War.* Washington, DC: Smithsonian, 1994.

FILMS

Almost from the beginning of American cinema, Native Americans and sport have attracted attention. Indeed, a handful of movies over the past century have centered on indigenous athletes, revealing popular attitudes toward Indians, sport, and society.

In the silent era, as two scholars contend, "While countless westerns in the silent film era were portraying the American Indian as a blood thirsty savage he fared better than any other minority group in the sports film of the era." Three films made during this period exemplify this pattern: *His Last Game* (1909), *Strongheart* (1914), and *Braveheart* (1925). *His Last Game*, set in Arizona, had cowboy hustlers trying unsuccessfully to bribe a principled pitcher of a Native American baseball team. *Braveheart*, a remake of *Strongheart*, centers on an Indian at a prestigious eastern university who hopes to use his education to lead a legal battle against a corporation determined to defraud a tribe of its fishing rights. In both *Braveheart* and *Strongheart* the Native American hero is a stellar athlete who takes the blame for a teammate and is expelled from school.

Life Begins in College, a 1937 Century-Fox production and a Ritz Brothers comedy, stars Nat Pendleton as a prosperous Native American football player who is brought off the bench and plays so spectacularly that he saves the coach's job. Importantly, Pendleton himself was a true athlete—an Olympic and then a professional wrestler—but had no Indian blood. This alarming exclusion is repeated and becomes canonical for modern sports cinema in America.

In 1951, *Jim Thorpe—All American,* also released as *Man of Bronze,* indisputably the most famous of all "Indian" sports movies, featured Euro-American actor Burt Lancaster as Thorpe. Despite Lancaster's memorable performance, the film has to be seen as an exploitative vehicle, with a white man assuming the role of one of the twentieth century's greatest athletes. It should be noted that supporting roles in this film were taken by Native American actors, such as Jack Big Head and Suni Warcloud.

Two decades later, in 1972, an indigenous athlete once again graced the silver screen. *When Legends Die* centers on a Native American protagonist in the world of rodeo. Red Dillon (Richard Widmark), an aging, hard-drinking rodeo cowboy, and his protégé, a Ute Indian Tom Black Bull (Frederic Forrest) enact a nuanced study of clashing cultures. Tom Black Bull has an amazing rapport with horses but eventually is forced to leave the reservation as the pressures of modernization and urbanization impact on his philosophy about land, space, legends and tribal loyalties. The bronco riding sections provide exciting and realistic cinema.

Following *Jim Thorpe–All American,* arguably the next most popular Native American movie with a sports scenario is the 1983 Buena Vista movie *Running Brave,* which recounts the true story of Oglala runner Billy Mills, who won the gold medal in the 10,000 meters at the 1964 Tokyo Olympics. Once again the starring role was played by a non-Native American, Robby Benson. The financing for the venture came from the Ermineskin Band of the Cree Indians. The story should have had wide popular appeal; sadly, lacking the backing of a major studio or a coast-to-coast release, the film was relegated to video markets.

Native American athletes have also played supporting roles in films. Notably, the Oscar-winning *One Flew Over the Cuckoo's Nest* (1975) features "The Chief," played by Creek actor Will Sampson. In a game between the patients of the mental hospital and the hospital staff, the Chief swats away scoring shots and scores a series of unstoppable points.

Finally, a number of documentary films have centered on Native Americans and sport. The *North American Society for Sport History Guide* (1993) lists twenty documentaries. Of particular interest are those concerned with indigenous sports, including *Lacrosse* and *Lacrosse: Little Brother of War.* The former shows lacrosse sticks being made by Mohawk Indians at a Cornwall factory, and how the Canadian Lacrosse Association promotes Native American involvement in lacrosse.

Scott A.G.M. Crawford

See also: Literature; Media Coverage; Scholarship.

FURTHER READING

Davidson, J.A., and D. Alder. *Sport on Film and Video: North American Society for Sport History Guide.* Metuchen, NJ: Scarecrow, 1993.

Zucker, M.Z, and L.J. Babich. *Sports Films: A Complete Reference.* Jefferson, NC: McFarland, 1987.

Sharon FIRTH

Born December 31, 1953, Aklavik, Northwest Territories, Canada
Cross-country skier

Sharon Firth, along with her twin sister Shirley, dominated North American cross-country skiing during the 1970s and 1980s. Firth, a Loucheux-Metis, born December 31, 1953, in Aklavik, NWT, Canada, was one of twelve children who grew up in a one-room house in Inuvik. She and her sister were the first athletes to represent Canada at four consecutive Olympic Games, from 1972 to 1984. Firth won nineteen Canadian Championship gold medals, the John Semmelink Memorial trophy for skiing excellence, was named to the Order of Canada in 1988 for her contributions to her country, and was inducted into the Canadian Ski Hall of Fame in 1990.

In 1969 Firth finished third in both the National Junior Championships five-kilometer race in Canada and the United

States. Firth was a member of the first Canadian women's team ever to participate in the World Nordic Championships, in Czechoslovakia in 1970. She achieved eleven victories in 1970, including the five-kilometer gold medal at the U.S. Junior Nationals and the ten-kilometer and relay at the Canadian nationals.

In 1972, after only five years of skiing, she was a member of the first Canadian women's cross-country ski team to attend an Olympic Games. Firth was the top Canadian in both the five kilometer, where she finished twenty-sixth, and the ten kilometer, where she placed twenty-fourth. In the relay, along with her sister Shirley, she helped the Canadian team finish in tenth position. After the games, she was the top North American at the Trans-Am Series between Canada and the United States. Following the Olympics, the Firth sisters became the first Nordic skiers to be awarded the John Semmelink Memorial Award by the Canadian Ski Association for their contribution to skiing in Canada.

In 1974, Sharon represented Canada at the World Championships in Sweden and then won gold in the North American championships in the ten kilometer, five kilometer, and relay. At the 1976 Olympic Games in Innsbruck, Austria, Sharon finished thirtieth in the five kilometer and twenty-eighth in the ten kilometer. The Canadian team finished seventh in the relay, the highest placing ever obtained by a Canadian cross-county ski team. Shortly before the Canadian Trials for the 1980 Olympics, the twins' mother was killed in a house fire. They decided nonetheless to ski and in qualifying for the 1980 Winter Olympics the Firths became the first North American female skiers to compete in three consecutive Olympics. Firth finished thirty-fifth in the five kilometer and helped the Canadian team finish eighth in the relay.

At the 1984 Olympic games, her twenty-first place finish was the best ever result by a Canadian in the twenty-kilometer race. She finished twenty-ninth in both the five and ten kilometer. Today the Firth Award, named for the twins, is presented to a woman who has made an outstanding contribution to cross-country skiing in Canada.

John Valentine

FURTHER READING

Bryden, Wendy. *Canada at the Olympic Winter Games.* Edmonton, ALTA: Hurtig, 1987.

Shirley FIRTH

Born December 31, 1953, Aklavik, Northwest Territories, Canada
Cross-country skier

Shirley Firth, along with her twin sister Sharon, dominated North American cross-country skiing during the 1970s and 1980s. Firth, a Loucheux-Metis, born December 31, 1953 in Aklavik, NWT, Canada, was one of twelve children who grew up in a one-room house in Inuvik. Along with her sister, she was the first athlete to represent Canada at four consecutive Olympic Games, from 1972 to 1984. Shirley won twenty-nine Canadian Championship gold medals, won the John Semmelink Memorial trophy for skiing excellence, was named to the Order of Canada in 1988 for her contributions to her country, and was inducted into the Canadian Ski Hall of Fame in 1990. Firth won the first race she ever entered and at fourteen won the United States five-kilometer Junior Ski Championship. She went on to sweep the medals at the next two U.S. Junior National Championships. She was a member of the first Canadian women's team to participate in the World Nordic Championships, in

Shirley Firth, and her twin sister Sharon, represented Canada in four consecutive Winter Olympics. *(AP/Wide World Photos)*

Czechoslovakia in 1970, where she was the top Canadian finishing in thirty-third place. Shirley achieved twenty-one victories in 1970, including the ten-kilometer gold medal at the U.S. Junior Nationals.

After only five years of skiing, at the age of eighteen, she represented Canada at the 1972 Olympic Games. Before the games in Sapporo, Firth contracted hepatitis and finished a disappointing thirty-fifth in the five kilometer. In the relay, along with her sister Sharon, she helped the Canadian team finish in tenth position, a best-ever finish for a Canadian team. After the games Firth finished as the second North American female (her sister Sharon was first) at the Trans-Am Series between Canada and the United States. Following the Olympics, the Firth sisters became the first Nordic skiers to be awarded the John Semmelink Memorial Award by the Canadian Ski Association for their contributions to skiing in Canada.

At the 1976 Olympic Games in Innsbruck, Austria, Firth finished twenty-eighth in the five kilometer and twenty-ninth in the ten kilometer. The Canadian team finished seventh in the relay, the highest placing ever obtained by a Canadian team. In 1978, Firth became the first woman ever to win all three titles at the Canadian National championships: the five, ten, and twenty kilometer. Shortly before the Canadian trials for the 1980 Olympics, the twins' mother was killed in a house fire. They decided to ski nonetheless and in qualifying for the 1980 Winter Olympics the Firths became the first North American female skiers to compete in three consecutive Olympics. Firth finished twenty-eighth in the five kilometer, twenty-fourth in the ten kilometer, and eighth in the relay. With her fourth-place finish in a 1981 World Cup race, Shirley recorded the best-ever finish by a Canadian. In 1982 she finished ranked eleventh in World Cup standings. Shirley married in 1983 and knew her fourth Olympic Winter Games would be her last. Her twenty-second-place finish in the ten-kilometer race was her best Olympic result ever. She finished twenty-eighth in the five kilometer and twenty-fifth in the twenty kilometer. Today the Firth award, created in 1985, is presented to a woman who has made an outstanding contribution to cross-country skiing in Canada.

John Valentine

FURTHER READING
Bryden, Wendy. *Canada at the Olympic Winter Games.* Edmonton, ALTA: Hurtig, 1987.

FOOTBALL

Some believe that American football may have roots in games indigenous to North America. Regardless of its origins, American Indians have contributed much to the sport as players, coaches, administrators, and symbols. To be sure, Native Americans had a more pronounced presence before 1945; nonetheless, they continue to be of vital importance to football.

The Formative Period

The rise of football as a modern sport coincided with the establishment of off-reservation boarding schools. It was on boarding school teams, particularly at Carlisle and Haskell, that Native Americans captured the public imagination and reshaped the game. Administrators and policy makers believed sports like football would facilitate the assimilation of indigenous individuals and communities; they did not anticipate the pride and excellence with which Native Americans would play the game. During the final decade of the nineteenth century and the first of the twentieth century, under the tutelage of Pop Warner, Carlisle would defeat traditional football powerhouses, including Harvard and the University of Chicago, and many of its players would distinguish themselves as All-Americans. Two decades later, Haskell Indian School would replicate this pattern. For progressive Indians at the time, such as Carlos Montezuma and Charles Eastman, football and the play of indigenous athletes held out the promise of attaining respect, equality, and self-determination.

Following their graduation from boarding schools, Native American athletes often played professional football. Early superstars included Jim Thorpe, Pete Calac, Joe Guyon, and Myles McLean. American Indians also were important as coaches and organizers of professional football. Indeed, Thorpe, recognized by many as one of the greatest athletes of the twentieth century, was the inaugural president of the National Football League.

Teams composed entirely of Native Americans were not limited to off-reservation boarding-school squads. As in other sports during the first third of the twentieth century, traveling squads and professional teams exclusively composed of Indians were formed. The Hominy Indians, a barnstorming team from Oklahoma featuring graduates from Haskell Indian School, toured the country in the 1920s, besting among others the world champion New York Giants. The Oorang Indians, composed of former Carlisle greats and coached by Jim Thorpe, was an original franchise of National Football League. Perhaps because of its lack of success, however, the team folded after only a few seasons.

Native Americans not only excelled on the field but on the sidelines as well. A number of early football standouts, including several graduates of Carlisle Indian School, went on to noteworthy careers as coaches. Albert Exendine, Egbert Ward, John Levi, and Tommy Yarr all turned to coaching after playing. Arguably the greatest Indian coach during the first half of the twentieth century was Lone Star Dietz, who led Washington State College (now Washington State University) to its only Rose Bowl victory before building Haskell into a powerhouse and finally

American Indians have played a prominent role in football. Here, powerhouses Carlisle Indian School and University of Chicago battle at Marshall Field in 1907. *(Courtesy of Chicago Historical Society)*

landing a job as head coach of the Boston (later Washington) Redskins.

Throughout this period, stereotypes about Indians were central to football. Even as the media celebrated the exploits of individual athletes, journalists often relied on racist notions in their coverage. Their accounts often centered on physicality, lack of discipline, poor training habits, supposed savage dispositions, interracial conflict, and tragic flaws. At the same time, misguided and romantic ideas about the frontier encouraged many schools and teams to adopt Native American mascots.

The Contemporary Period

The indigenous presence in football declined after 1930. The end of the off-reservation boarding school system and the later desegregation of collegiate and professional sports were two of the more important factors contributing to the reduced participation of Native Americans in football. Of course, American Indians continued to watch, play, and coach football at all levels of the game.

The number of American Indians who received national recognition as players or coaches noticeably dropped in the last two-thirds of the twentieth century, and collegiate and professional teams composed entirely of indigenous talent disappeared from American playing fields. However, native superstars have continued, if more intermittently, to delight audiences. Noteworthy players have included Jim Plunkett, Sonny Sixkiller, Wahoo McDaniel, and Jack Jacobs.

As during the formative period, Native Americans have played often unrecognized organizational and administrative roles. Few know that the Houston Oilers, one of the founding franchises of the American Football League (now relocated to Nashville and renamed the Tennessee Titans), has for its entire history been owned by a Cherokee, Bud Adams.

In the contemporary period, even as Indian players and coaches have become more marginal, Indian imagery has remained central to the sport. Importantly, Native American mascots have become contentious. Thus, while numerous football teams still profit from images of Indians, an increasing number (including Stanford University and Dartmouth College) have opted, in response to indigenous activists, to end their reliance on imaginary Indians.

C. Richard King

See also: Carlisle Indian Industrial School; Haskell Institute; Hominy Indians; Mascot Controversy; Oorang Indians.

FURTHER READING

Montezuma, Carlos. "Football as an Indian Educator." *Red Man* (January 1900): 8.

Oriard, Michael. *Reading Football*. Chapel Hill: University of North Carolina Press, 1993.

Oxendine, Joseph B. *American Indian Sports Heritage*. 2d ed. Lincoln: University of Nebraska Press, 1995.

Sculle, Keith A. "'The New Carlisle of the West': Haskell Institute and Big-Time Sports, 1920–1932." *Kansas History* 17 (1994): 192–208.

FORT MICHILIMACKINAC

In 1763, a group of Native Americans attacked British soldiers at Fort Michilimackinac under the guise of a lacrosse game, as part of the larger Indian uprising against the British, who had taken over French lands in Canada and the Great Lakes Region at the conclusion of the French and Indian War (1754–1963). *Baggatiway*, or lacrosse, was the favorite team game of eastern Indians. Lacrosse game as a war strategy was used against the British at only one of the various attacks on fortifications from Forts Niagara and Pitt to Detroit and Sandusky. Collectively these attacks became known as Pontiac's Rebellion (or War), after Chief Pontiac of the Ottawa Indians.

One of the fortifications was Fort Michilimackinac, located between Lake Michigan and Lake Huron. The Ojibwe Indians conspired to defeat its English garrison of ninety men. On June 4, 1763, King George III's birthday, nearly 400 Indians were outside the fort playing *baggatiway,* what the French called *le jeu de la crosse.* This game of "much noise and violence," according to Alexander Henry, who was at the scene, diverted the English soldiers. The Indians threw the ball into the fort and retrieved it several times, as a diversionary tactic. Supporters in attendance hid weapons under their garments. When the attack was made, many of the soldiers were killed, and a score were taken to Montreal and ransomed at high prices. This was an early North American case of the use of sport for political and military goals. Four months later, a proclamation by King George closed the British frontier west of the Appalachian Mountains.

Ronald A. Smith

See also: Lacrosse.

FURTHER READING

Armour, David A., ed. *Massacre at Mackinac—1763: Alexander Henry's Travels and Adventures in Canada and the Indian Territories Between the Years 1760 and 1764*. Macinack Island, MI: Mackinac Island State Park Commission, 1966.

Schmalz, Peter S. *The Objibwa of Southern Ontario*. Toronto: University of Toronto Press, 1991.

A great goaltender, Grant Fuhr was integral to the Edmonton Oilers dynasty of the 1980s. *(AP/Wide World Photos)*

Grant Scott FUHR

Born September 28, 1962, Spruce Grove, Alberta, Canada
Hockey player

Grant Fuhr was one of the greatest goaltenders in the history of the National Hockey League. A key component of the dynamic Edmonton Oilers dynasty of the 1980s (which was led by Wayne Gretzky), Fuhr won five Stanley Cup rings during his nineteen-year NHL career. By the time he retired at the end of the 1999–2000 season, he had won 403 regular season games—a mark reached by only a handful of netminders.

Fuhr was born in Spruce Grove, Alberta, on September 28, 1962. He never knew his parents, who were unmarried teenagers not mature enough to raise a child. His father was black, and his thirteen-year-old mother was likely part Cree. Fuhr's adoptive parents encouraged their son to hone his athletic abilities. He passed up an opportunity to play professional baseball in the Pittsburgh Pirates organization in order to concentrate on hockey. After playing two years in the developmental Western Hockey League, Fuhr broke into the NHL with the Oilers in 1981 at the age of eighteen.

Fuhr's career can be divided easily into two distinct parts. The first phase (1981–1991) included Stanley Cups, six All-Star Games, and the Verzina Trophy as the NHL's top goaltender in 1988. But he earned a reputation for irresponsible behavior when not on the ice. He married and divorced twice during this era. In 1990 the league suspended him for six months for habitual cocaine use, and the Oilers traded him at the end of the 1990–1991 season.

The second phase of Fuhr's career (1991–2000) was marked by several ups and downs. Although he overcame his drug use, he suffered through losing seasons, was traded often, and occasionally found himself a backup to another goaltender. But he persevered to emerge as one of the most durable players in league history. He set an NHL record in 1995–1996 by playing in seventy-nine games—the most ever by a goaltender—as a member of the Saint Louis Blues. Ultimately, the second half of his career proved to be successful, even if not as spectacular and award-filled as the first half.

Drugs were not the only problem Fuhr had to deal with as a professional hockey player. Like all such athletes who play several years, he was operated on for a variety of injuries. Reconstructive surgery on his knees and left shoulder allowed him to play long enough to achieve hockey immortality. Observers of the sport are practically unanimous in their belief that he will win a place in the NHL Hall of Fame. Moreover, retirement from hockey did not mean the end of athletic endeavor for Fuhr, as he hopes to compete on the Canadian professional golf tour.

Roger D. Hardaway

FURTHER READING

Murphy, Austin. "Old Faithful." *Sports Illustrated,* February 19, 1996.

Wiley, Ralph. "The Puck Stops Here." *Sports Illustrated,* January 11, 1988.

Steve GACHUPIN

Born September 1, 1942
Mountain runner

Six-time Pikes Peak Marathon winner and 1968 Olympic time trial participant Steve Gachupin always says he runs "to bring honor to his village."

Born in 1942, Gachupin became one of the greatest mountain runners in history. As a boy growing up at Jemez Pueblo in New Mexico, he ran with his cousin, Sefredo Toledo, through the surrounding mountains every morning. At Jemez people have traditionally run for ceremonial reasons, as a time to pray, as a mode of message carrying, and for transportation. The long history of running at Jemez and Gachupin's morning mountain-running regime helped prepare him to become one of his people's greatest runners and an ambassador for the sport as well as of the Pueblo.

In July 1966, having won the ceremonial races at his Pueblo for five consecutive fall seasons, Gachupin decided to participate in a race being held in Denver. His trip to Colorado was his first out-of-state journey, and was the first time he had ever ran in a marathon. Gachupin finished second, taking no water or aid throughout the competition. His stamina and untrained talent impressed race organizers and reporters, who encouraged him to participate in the race to the top of Pikes Peak and back the following month. That August, Gachupin raced to the 14,110-foot summit and back to win his first marathon.

Winning the Pikes Peak Marathon annually from 1966 to 1971 and becoming the first man to run the entire distance earned Gachupin the title "King of the Mountain." However, Pikes was not the only mountain he conquered. He was also champion five times at Sandia Peak on the La Luz Trail run and raced in more than forty different marathons around the country throughout his career. The U.S Congress and *Sports Illustrated* have recognized his outstanding running; however, his champion status comes, according to three-time Olympic track and field coach Joe Vigil, from "his humility and longevity in the sport. A lot of people quit running after they stop winning, but Steve always brings his family and people from Jemez and continues to run in events all around the region. Steve represents the sport the way it should be represented."

Gachupin considers one of his greatest accomplishments to be the runners he himself has trained. As coach of the Jemez Valley High School Cross Country team he has led the school to an unprecedented fourteen state championships. He teaches his runners not only to succeed in running but also to "achieve their educational goals. By being a good sportsman they can earn their education, health and wellness, and togetherness in their community and among all people." His dedication to running and coaching earned him the Runner's World Golden Shoe Award in 1992 and a 2001 New Mexico High School Coaches' Association award. As a runner and community leader he has supported events that ceremonially reenact the Messenger Run of the 1680 Pueblo Revolt, races to support local schools, WINGS of America events, and runs to garner support for issues that affect the reservation and Native American people.

Steve Gachupin is still an active runner and community leader. Leading by example through his own running and through his support of younger runners, he continues to succeed at his humble goal, "to bring honor to my village."

Brian S. Collier

FURTHER READING

"One Very Fast Man and His Mountain." *Colorado Springs Gazette* (www.skyrunner.com/story/2001p2.htm), August 16, 2001.

GAMBLING

Today when one thinks of gambling and Native Americans, what springs to most people's minds are casinos on Indian reservations. However, gambling has long been an integral part of Native American societies. Gambling plays a prominent part in many traditional Indian games and sports. Therefore, it is no surprise that today Indian casinos are proliferating throughout the nation.

Gambling was widespread throughout Native American culture prior to contact with the white man. Indians wagered on a variety of games of chance and skill. Indeed, most of the games mentioned in this volume had gambling as an important component. Items wagered included clothing, weapons, horses, and implements used in the games themselves. In extreme circumstances, wives and children served as stakes. Gambling was a vital part of Indian games and an important component of Native American culture. Gambling has had great social, economic, political, spiritual, religious, and mythological importance for Native Americans.

Gambling served as one of the primary means of facilitating the movement of goods between tribes. An eyewitness account of a Sing Gamble held on the Puyallup Reservation in Washington State in 1895 illustrates the role of Native American gambling. The Sing Gamble featured tribe members singing and chanting to the booming beat of drums as two teams of players

Gambling, long an important element of indigenous athletics, has become key to the economic development of many native communities in the past decade as casinos like Foxwoods Resort Casino, located on the Mashantucket Pequot Indian Reservation in Connecticut, have proliferated. *(AP/Wide World Photos)*

participated in a tense game akin to the shell game in white culture. Each tribe placed items in the pot of booty, which would be the winners' prize. A dealer shuffled a set of ten wooden chips; one chip marked differently from the rest was the players' quarry. The dealer hid each chip under one of two piles of wood shavings; each team took turns guessing the location of the elusive chip until one team correctly guessed where it rested. The winning team took possession of the pot. The game continued until the visiting Black River tribe had lost all the items it had brought to wager, including the chief's horse.

Additionally gambling was the preferred means by which many tribes exchanged goods in lieu of raiding or open warfare. Gambling's cultural significance, however, was of far more consequence than its material function. For instance, the Hurons played a dice game to heal the sick. Many of the more intense sessions of wagering occurred at festivals of great religious and spiritual import to the tribes. Games and gambling were important features of Native American religious/mythological stories. In many cases, these myths were the sources of Native American games and rules of play. Many tribes believed that playing these games brought the individual closer to the gods or prepared one to interact with the gods in the afterlife.

Gambling stories predominate throughout the continent, but according to Gabriel, they contain some common themes. These myths often include a good gambler (a god) and an evil gambler, or trickster (another god). The person victimized by the evil god is often an ordinary person who lives next door to him. The evil entity entices the ordinary person into irresistible games of chance. The victim loses everything, including family and tribe members, until he is naked and destitute. Sometimes the victim is sold into slavery or loses body parts. However, the good gambler redeems the victim through cunning, skill, or magic, restoring the victim's health and possessions and visiting the same miseries upon the vanquished evil gambler. The good gambler is often the tribal hero and may appear in human or animal shape or in the form of a natural process. The good gambler's victory is often portended by some natural or cultural phenomenon, such as fortuitous rains, a good harvest, or a successful hunt.

Tribal reservations have long been among the bleakest areas in America. Reservations have long suffered from poverty, lack of adequate health care and education, high rates of crime, alcohol and drug abuse, and mental health problems. Tribes desperate for ways to revitalize their economic and social well-being turned to gambling, long a staple of Native American life.

Before the advent of casino gambling, Native American reservations featured bingo games and poker machines to entice tribal members and non-Native American tourists into gambling on Indian lands. The Penobscot tribe of Maine was the first tribe to open a bingo hall, in 1976. The Penobscot ran their weekly Sunday night games until 1980. However, the Seminole of Florida spurred the boom of high-stakes gaming. In 1979, the Seminole and the Cabazon of California were among the first tribes to plunge into the high-stakes gaming business with the opening of high-stakes bingo halls that offered prizes that exceeded the maximum cash prize allowed by state law. The states of Florida and California threatened to shut down the burgeoning gaming operations; however, the tribes sued in federal court. The tribes won the landmark *California v Cabazan Band of Mission Indians* (480 U.S. 202 1987) and earlier, in a lower court, *Seminole Tribe v Butterworth* (658 F.2d 310 1981). The U.S. Supreme Court ruled that by treaty, the reservations were sovereign nations in domestic matters, that gaming fell under this power, and that therefore the tribes were able to establish their own gaming entities. The Cabazan decision led to a proliferation of tribal casinos and bingo halls across the continent.

The next phase in the history of Indian gambling establishment was heralded by the passage of the Indian Gaming Regulatory Act of 1988 (100 P.L. 497; 102 Stat. 2467). This goal of this act was to regulate Native American gaming in order to promote tribal economic growth, self-sufficiency and self-government, and to guard against organized crime's penetration of tribal gambling operations. The law stated that gambling establishments on tribal lands were to be owned by the whole tribe; however, tribes could contract with outside entities to run the casinos or bingo halls. Further individual tribe members were prohibited from owning gaming establishments. To oversee the law and to regulate Indian gaming, the act established the National Indian Gaming Commission. The primary purpose of the commission is threefold: to shield tribes from the influence of organized crime, to make certain that tribes are the primary beneficiaries of revenue generated by gambling establishments, and to ensure that all gaming is conducted fairly by both operators and players.

In 2001, two hundred tribes had stakes in 290 gambling establishments generating revenue of $12,735,379. While some tribes have ameliorated some of the worst poverty and problems on their reservations, many tribes have found gambling is not the panacea they hoped. And yet, the Puyallup have taken the plunge, with a casino and bingo hall on their reservation.

Rick Dyson

FURTHER READING

Gabriel, Kathryn. *Gambler Way: Indian Gaming in Mythology, History and Archaeology in North America.* Boulder, CO: Johnson Books, 1996.

Lane, Ambrose I. *Return of the Buffalo: The Story Behind America's Indian Gaming Explosion.* Westport, CT: Bergin and Garvey, 1995.

National Indian Gaming Association (www.indiangaming.org).

National Indian Gaming Commission (www.nigc.gov).

Stein, Wayne J. "Gaming: the Apex of a Long Struggle." *Wicazo Sa Review* 13, no. 1 (1998): 73–91.

Jacob Gill "Jake" GAUDAUR

Born April 4, 1858, Orillia, Ontario
Died October 11, 1937, Orillia, Ontario
Sculler

Jake Gaudaur was Canada's greatest sculler of the late nineteenth century and one of the best in the world during the heyday of professional sculling. His powerful physique and commitment to physical conditioning endowed him with exceptional strength and stamina. This big, quiet, dignified man won over 200 races and many titles during his career, including three consecutive U.S. titles in the 1880s and the world title in 1896. He also held the long-unbroken record of racing three miles in 19:01.5 at Austin, Texas, in 1894.

Gaudaur was born on April 4, 1858, in Orillia, Ontario. He was of French-Canadian, Scottish, and Ojibwe decent, his paternal grandmother being the daughter of a local Ojibwe chief. He grew up in the Narrows District between Lakes Couchiching and Simcoe and learned to row at an early age. He also formed his deep love of the outdoors here and developed his skills as a fishing guide from his lifelong association with the local Ojibwe. In 1876, when he was only eighteen, his friends encouraged him to enter the Orillia Regatta, which used rowboats. In this, his first race, he easily beat his opponents. He later entered another race at Barrie and won that as well. For the next four years, he competed almost entirely in Canada in singles and doubles races using lapstreak skiffs.

In 1879, Gaudaur sought to establish himself as a real force in the rowing scene and began competing in professional races using a single shell, contending against such well-known scullers as the American Fred Plaisted and fellow Canadian Ned Hanlan. He surprised many by finishing third in the Barrie Annual Regatta. Impressed with Gaudaur's abilities, Hanlan guided and encouraged the young sculler. Under Hanlan's coaching, Gaudaur's skills grew steadily, and by the early 1880s he was regarded as one of Canada's most promising young scullers.

Gaudaur moved to St. Louis in 1882 to improve his chances of gaining financial backing so he could enter more races. Now that he was competing regularly in top races, his successes increased in number; in 1885 he took eight of the eleven major races he competed in. By this time, he was considered to be in the front rank of scullers worldwide.

Over the next decade, Gaudaur enjoyed unparalleled success. He won the American Sculling Championship on three consecutive occasions (1885–1887). In 1894 at Austin, Texas, he raced three miles with a turn in 19:01.5, a world record. He attained his highest ambition on September 7, 1896, when he decisively defeated the Australian James Stanbury in London for the world championship. Gaudaur retained the title for another five years, relinquishing it when defeated by George Towns in 1901. Now well past forty, Gaudaur announced his retirement from competitive rowing.

In retirement, Gaudaur engaged in several successful business endeavors, running hotels and outfitting, boat rental, and fishing operations. As a fishing guide, he was much sought after, partly owing to his fame as an athlete but mostly because of his skill as an outdoorsman. He continued to row for his own recreation and enjoyed many honors and tributes throughout his retirement. Gaudaur married three times and had eleven children, six by his second wife, Ida, and five by his third wife, Alice. He died in Orillia of leukemia on October 11, 1937.

Edward W. Hathaway

FURTHER READING

Brooks, Stephanie Anne. *An Athlete Biography of a Champion Canadian Sculler: Jacob Gill Gaudaur, 1858–1937.* Master's thesis, University of Western Ontario, 1981.

Robert GAWBOY

Born June 28, 1932, Vermillion Lake Indian Reservation
Died July 15, 1987
Swimmer

A leading high school and college swimmer, Robert Gawboy held national high school records, placed high in national collegiate competitions, and set a world record of 2:38 in the 220-yard breast stroke at the National AAU Indoor Swimming championships in 1955. Gawboy was small of stature but was powerfully built and had great determination to succeed, despite ailments that ultimately shortened his career.

Gawboy was born on June 28, 1932, on the Vermillion Lake Indian Reservation to Robert Gawboy, an Ojibwe, and Helmi Jarui, a white of Finnish decent. His family moved to Ely, Minnesota, when he was fourteen. He competed on the Ely High School swim team, winning the state championship in 1949. That year, Gawboy won the 100-yard breast stroke in 1:08.3. The following year, as a senior, he broke the national high school records for the 120-yard

Robert Gawboy broke the world's record in the 220 yard breast stroke in 1955. *(AP/Wide World Photos)*

individual medley with 1:14.8, and the 150-yard individual medley with 1:38.0.

Gawboy first showed signs of the multiple sclerosis that later incapacitated him shortly after finishing high school, when he noticed trouble with his coordination. This soon cleared up, however, after he enrolled at Purdue. He made the swim team and placed first in the 150-yard individual medley in an East-West collegiate competition. In the 1952 NCAA National Championships, he placed second in the 150-yard individual medley. A congenital blood condition in his left leg (arteriorvenous fistula) coupled with a recurrence of coordination problems and declining endurance, forced him to quit swimming. He underwent surgery in December 1954, however, and was able to begin working out again in February 1955.

By this time, Gawboy was enrolled on a scholarship at the University of Minnesota. He competed in the AAU Indoor National Champions at Yale in April, 1955, his first race in over two years. Despite this long hiatus and only five or six weeks of training, he surprised everyone by breaking the American record time in the preliminaries for the 220-yard breast stroke, then going on to capture the finals and set a new world-record time of 2:38.0.

Shortly afterward his circulatory problems recurred. He was operated on again the following December, but this time it did not help. Newly married and doing poorly in school, he decided to retire from swimming. Several years later Gawboy returned to aquatics, first joining a water polo club in 1966 and later competing in the Senior Men's division of the AAU, swimming in the 1967 State Outdoor Championships. His difficulties with coordination worsened and he was forced to give up swimming competition entirely. In 1969 he was diagnosed with multiple sclerosis. The progress of the disease accelerated, and he was forced to retire on disability in 1972. In 1980, he was inducted into the American Indian Athletic Hall of Fame in Lawrence, Kansas. He died on July 15, 1987.

Edward W. Hathaway

FURTHER READING

"Medical Ailments Robbed Bob Gawboy." *Minnesota Swimmer*, February 15, 1973.

Werden, Lincoln A. "Gawboy, Jack Wardrop Lower World Marks in U.S. Swim Meet." *New York Times*, April 2, 1955.

GENDER RELATIONS

In Euroamerican sport, gender equity policies and other measures try to ensure equal or equitable opportunities in sport involvement for both women and men. Such policies often rely on Euroamerican conceptions of gender relations, which often do not take non-Euroamerican cultural aspects of sport participation into consideration. Nevertheless, "First Nations sport culture . . . reflect[s] ongoing shifts in gender relations between Native and non-Native communities, and within Native communities." As a result, sporting festivals have had to cope with the challenge of gender equity, as evidenced by changes in the Arctic Winter Games, Northern Games, and competition powwows.

Arctic Winter Games

The Arctic Winter Games (AWG), which began in 1970, currently has two categories for indigenous sport: Arctic sports became an official category in 1974, while Dene games (traditional indigenous competitions) were added in 1990. However, women's categories for Arctic sports were not added until 1982, while women's categories for Dene games are only proposals for the 2004 Games. Some people argue forcefully that women traditionally did not play many Inuit and Dene games and thus should not compete in them at the Arctic Winter Games. Such arguments have been undermined by changing gender roles and a drive toward gender equity in general.

Northern Games

The Northern Games are an Inuit sporting/cultural festival that started in 1970. For the first twenty years of the Northern Games, the only women's event was the "Good Woman Contest." According to the 1975 Northern Games program, "the Good Woman competition has been a unique and very popular part of the Games from the beginning. Women from many settlements have shown their skills in sewing, cooking, preparing food and skins, and dancing—all the things necessary to be a good wife and mother. Traditionally looks were not the prime asset for finding a good woman but rather her ability to contribute positively to family and community." Good Woman Contest events included fire building, tea boiling, bannock making, fish boiling, fish cutting, seal skinning, muskrat skinning, duck plucking, sewing, jigging, and traditional dress.

Men's events, which like the Good Woman Contest events varied from year to year, required a greater degree of athleticism. Some of the men's events were the one-foot high kick, two-foot high kick, Alaskan high kick, knuckle hop, ear weight, and arm pull. Many of the games from 1976 on had a men's category in the Good Woman Contest, wherein men would compete in the same events as the women. The event is conducted in a somewhat facetious manner: "While men on the traplines do carry out many of the tasks included in the Good Woman Contest, their performance in this event often generates laughter from the audience more so than it does amazement over their skills." Women, however, did not start competing in the traditionally male events until the late 1980s and early 1990s.

Competition Powwows

Gender lines are clearly drawn and carefully maintained at competition powwows. Powwows call on gender scripts and reinforce gendered norms. The commencement of a powwow is marked by the Grand Entry, where adult males enter the arena as a group, followed in order by the adult females, boys, and finally girls. Competitors are divided not only by age but also by sex, with women competing in different events than men. Like Inuit and Dene games, women's full participation in powwow dancing began after men's. Originally, women had a supportive dance role, standing behind the drum. Women now dance in the main area and have adopted a style of dance that is more vigorous and thus more closely resembles that of men.

Menstruation has a large impact on women's participation in competition powwows, as women are not supposed to dance while they are menstruating. Two types of explanations have been offered for practice: impurity and enhanced power. The impurity hypothesis views menstruation as a time where women can pollute those around them, bringing bad luck. This hypothesis has been rejected by many Aboriginal peoples, who instead view menstruation as a time of enhanced power. In either case, menstruating women are typically supposed to be separated from men at powwows and thus have fewer opportunities to dance in competition powwows.

Audrey R. Giles

See also: Arctic Winter Games; Competition Powwows; North American Indigenous Games.

FURTHER READING

Anderson, K. *A Recognition of Being.* Toronto: Second Story, 2000.

Heine, M. *Dene Games: A Culture and Resource Manual.* Yellowknife, NWT: Sport North Federation and MACA (GNWT), 1999.

———. "The Symbolic Capital of Honor: Gambling Games and the Social Construction of Gender in Tlingit Indian Culture." *Play & Culture* 4 (1991): 346–58.

Mandelbaum, D.G. *The Plains Cree: An Ethnographic, Historical and Comparative Study.* Regina: Canadian Plains Research Centre, 1979.

Northern Games Association. *Northern Games 1976 Program.* Inuvik, NWT: Northern Games Association, 1976.

Paraschak, V. "Doing Sport, Doing Gender." In *Sport and Gender in Canada,* edited by Phillip White and Kevin Young. New York: Oxford University Press, 1999.

———. "Variations in Race Relations: Sporting Events for Native Peoples in Canada." *Sociology of Sport Journal* 14 (1997): 1–21.

Roberts, C. *Pow Wow Country.* Helena, MT: American and World Geographic, 1992.

White, J.C. *The Powwow Trail: Understanding and Enjoying the Native American Pow Wow.* Summertown, TN: Book Publishing, 1996.

Angus Simon GEORGE

Born 1910, St. Regis, Quebec, Canada (Akwasasne Mohawk Reserve)
Died January 8, 1992, St. Regis, Quebec, Canada
Lacrosse player

George, a Mohawk, grew up on the Akwasasne Reserve. He attended school locally at Glen Nevis playing hockey and lacrosse. By his late teens he was recognized as one of the leading lacrosse players on the reserve. In the early 1930s he was playing with the Cornwell Island Indians in a league with other area towns. In the middle of the decade he left to play professional box lacrosse for teams in several North American cities, including Syracuse, Wilkes-Barre, Vancouver, and Los Angeles. He returned to Quebec in 1940, playing for a combined Kahnawake-Akwasasne team in the Quebec league. This team dominated the league for much of the decade. In 1962 George played his last game, an exhibition match in Quebec. He was a construction ironworker after his retirement from sports, working on the St. Lawrence Seaway, among other projects. He enjoyed many tributes and honors later in life, including the naming of a street in St. Regis after him and induction into the Glengarry Sports Hall of Fame in 1995.

Edward W. Hathaway

FURTHER READING
Glengarry Sports Hall of Fame, archives.

James W. GLADD

Born 1922
Died 1977
Baseball player

Jim Gladd, a mixed-blood Cherokee, had a brief major league baseball career in 1946. During 1947–1955 he was a catcher for four franchises in the Pacific Coast League.

After graduating from a Fort Gibson, Oklahoma, high school in May 1940, Gladd was signed by a scout for the New York Giants of the National League. Assigned to Milford, Delaware, of the Class D Eastern Shore League, he hit .218 with nine homers and forty-one runs batted in (RBIs) during the 1940 season.

In 1941, Gladd was selected as the catcher on the All-Star team of the Class D North Carolina State League. Playing for the Salisbury Giants, he hit .248 with seven homers and forty-eight RBIs. On November 1, 1941, the Jersey City Giants of the Class AA International League acquired Gladd's minor league contract. After playing briefly for Jersey City, New Jersey, and batting his career-best .286 at the beginning of the 1942 season, he was assigned to Fort Smith, Arkansas, of the Class C Western Association for more experience. During the remainder of the 1942 season at Fort Smith, he hit .253 with no homers and fourteen RBIs.

Gladd served in the military for the 1943–1945 seasons. A strong showing in spring training in 1946 earned him a return to Jersey City. Batting .269 with three homers and twelve RBIs, he was called up to the parent New York Giants of the National League. Gladd made his major league debut on September 9, 1946, in a game against the Brooklyn Dodgers. Catching four games that month, he made no errors, but had only one hit in eleven at bats for a .094 batting average.

With veteran catchers Walker Cooper and Ernie Lombardi on their roster, the New York Giants optioned Gladd to the San Francisco Seals of the Class AAA Pacific Coast League for the 1947 season. With the Seals, Gladd batted .252 with five homers and thirty-three RBIs. Optioned to the Hollywood Stars of the Pacific Coast League for the 1948 season, he batted .246 with seven homers and thirty-five RBIs.

Still the property of the New York Giants, he was assigned outright to the Portland (Oregon) Beavers of the Pacific Coast League on January 24, 1949. With Portland in 1949 he hit .268 with nine homers and his career-high fifty-seven RBIs. Returning to Portland in 1950, Gladd hit .267 with fifty-five RBIs and a career-high sixteen homers.

Called back to active military service, he served in Korea and missed the entire 1951 season. Returning to Portland for the 1952–1954 seasons, his best efforts were a .282 batting average, three homers, and forty-nine RBIs in 1953. Batting .227 for the season, Gladd concluded his baseball career with the San Diego Padres of the Pacific Coast League on May 12, 1955.

Royse Parr

FURTHER READING
Johnson, Lloyd, and Miles Wolff. *The Encyclopedia of Minor League Baseball.* Durham, NC: Baseball America, 1999.
Professional Baseball Player Database Version 4.1 (1930–1985). Shawnee Mission, KS: Old Time-Data.
Unpublished baseball contract cards for James W. Gladd. National Baseball Hall of Fame and Museum. Cooperstown, New York.

GOLF

Although Native Americans have played golf since the sport's earliest days in the United States, Indian golfers did not make a national impact until the last decades of the twentieth century. For most of the 1900s, restrictive racial policies of all-white organizations like the Professional Golfers Association (PGA) and the U.S. Golf Association (USGA) limited opportunities for Native Americans. When the civil rights movements of the 1950s and 1960s removed racial barriers to participation, Indian golfers found success on the professional tour. Finally, in the 1990s, the Native American community built national golfing organizations to encourage Indian golfers further.

The Native American presence in golf dates back at least to the 1891 construction of Shinnecock Hills (Long Island) Golf Club, one of America's first modern courses. The club was carved out of a former Indian reservation, and local Shinnecock Indians provided the bulk of the labor. Some of the Native American workers learned the game, and when Shinnecock Hills hosted the second U.S Open Tournament in 1896, one Indian, Oscar Bunn, participated in the event.

Bunn joined John Shippen, the son of an African-American missionary to the Shinnecock, as the only two native-born players in the event. When the English and Scottish professionals, who accounted for most of the entries, learned of the two nonwhite competitors, they threatened a boycott. Theodore Havemeyer, a wealthy sugar refiner and the first president of the USGA, stood by Bunn and Shippen and refused to heed the segregationists' ploy. The white golfers sheepishly dropped their protest, and the Open proceeded as scheduled. Bunn finished out of the lead; his subsequent career is unknown. John Shippen finished fifth in 1896 and thereafter competed in several more U.S. Opens. The American golfing establishment, however, did not build upon Havemeyer's inclusiveness.

The Jim Crow Era

By the 1920s, the USGA, the PGA, founded in 1916, and most private country clubs actively discriminated against nonwhite golfers. The PGA most infamously demonstrated this attitude in 1943 when it added a "Caucasian clause" to the requirements for membership in its constitution. In this prejudicial atmosphere, nonwhite golfers built parallel structures to support their own participation. The United Golfers Association (UGA), founded in 1926, was the most successful of these organizations. Although primarily run by and for African Americans, the UGA accepted all players, including Native Americans, who were denied access to the white golfing world.

Bill Spiller, who was of mixed African-American and Cherokee heritage, was the most successful Indian golfer in the UGA, and he won numerous tournaments in the 1940s and 1950s. More importantly, Spiller, along with black golfers Ted Rhodes and Charlie Sifford, campaigned strenuously against the discriminatory practices of the PGA. Spiller sued the organization in 1948 and throughout his career marshaled public opinion against the "Caucasian Clause." His battle ended in 1961 when the PGA, under intense public pressure, dropped its racial membership requirements. The victory came too late for Spiller, who was past his prime as a golfer.

Increasing Indian Participation

Several black golfers, like Charlie Sifford, made an immediate impact when the PGA finally opened its doors. Native Americans, however, had little institutional support and were slower to achieve national success. Over the next twenty years, a few self-taught Indian golfers made their mark. In 1969, Orville Moody, a Choctaw who had won the 1958 All-Army golf championship, became the first Native American to play in the U.S. Open since Oscar Bunn. A virtual unknown, he was a local qualifier and shocked the golfing world by winning the event. Five years later Rod Curl, a Winta, became the second Native American to win a tour event, at the 1974 Colonial National Invitational. Few other Indian golfers enjoyed similar success.

In the 1990s, led by the phenomenal success and notoriety of Tiger Woods, who is part Native American, the sport of golf shed its elitist reputation, as new groups of people embraced the sport. Native American golfers, for example, banded together to create a community of support by sponsoring numerous all-Indian golf tournaments, like the Olgala National Open (competed since 1986), the Gallup Indian Ceremonial Golf Tournament (inaugurated in 1999), and the Native American Golf Tour Pro-Am (founded in 2002). Furthermore, since 1996 the USGA has been a financial partner of the Native American Sports Council, and the two organizations have worked to build a strong institutional framework to encourage aspiring Native American golfers.

This renewed Indian interest promises to change the face of American golf yet further. In 1999, Stanford-educated Notah Begay became the first Native American PGA tour winner since Rod Curl and knocked down another barrier as the first Indian to compete in the Master's in Augusta, Georgia. In the first years of the twenty-first century Begay was the only Indian on the professional tour, but the groundwork has been laid for a new generation of Native American golfers to join him.

Gregory Bond

FURTHER READING

Dawkins, Marvin P., and Graham C. Kinloch. *African-American Golfers During the Jim Crow Era.* Westport, CT: Praeger, 2000.

Oxendine, Joseph B. *American Indian Sports Heritage.* Champaign, IL: Human Kinetics Books, 1988.

GOVERNMENT PROGRAMS AND INITIATIVES—CANADA

Funding for sport in Canada occurs at the federal, provincial/territorial, and municipal levels. An agreement between the federal and provincial/territorial governments reached in 1985 resulted in a clarification of federal and provincial/territorial roles and jurisdictions. Municipal governments are primarily responsible for facility development. Provincial/territorial governments in Canada are responsible for funding, sport development, supporting provincial/territorial championships and other games, and developing facilities and supporting athletes and their respective provincial or territorial sport organization. The federal government's role at the national level essentially parallels that of provincial/territorial governments at the provincial/territorial level. The federal government's focus is on funding for high-performance sport, sport development, supporting national championships and major games (e.g., the Olympics), and supporting athletes and their respective national sport organizations.

While sport initiatives and programs are advertised as open to all people, the federal government has acknowledged that Aboriginal peoples are underrepresented in the Canadian sport system. In an effort to address the issue of underrepresentation, in the early 1970s the federal government introduced a sport programming initiative that funded Aboriginal associations in five provinces. The goal of this initiative was to provide increased opportunities for Aboriginal participation in sport in order that Aboriginal people could compete in mainstream sport events. Total funding for this initial two-year program (1970–1972) was $325,000.

In 1970, the federal government also commenced funding of the Arctic Winter Games. The initial federal contribution was approximately $250,000. The federal government has continued to support these biennial games through various agencies and departments. Additional Canadian support is provided by the provincial government of Alberta and the territorial governments of the Northwest Territories and the Yukon. Some government support for team travel is provided when the games are hosted outside of Canada.

In 1973, the federal government commenced funding a five-year experimental program known as the Native Sport and Recreation Program. It had an initial run from 1973 to 1978. This program built on the initial two-year program introduced in 1970. The five-year program was extended a further two years, from 1978 to 1980. Approved funding for the extension was $2.79 million. A further one-year extension was contemplated, and one million dollars were allocated. Continued funding came with increased pressure by the federal government for Aboriginal athletes to assimilate into the mainstream sport system. A number of Aboriginal peoples resisted and asked the government to support local Aboriginal games. The federal government objected, and funding to the Native Sport and Recreation Program was cut in 1980. Following the elimination of this program, the federal government made no further effort to develop sport programming in Aboriginal communities.

A series of reports criticizing the absence of opportunities for Aboriginal people in sport and recreation resulted in the federal government's contributing monies to several different Aboriginal sport initiatives. In 1994 the federal government commenced funding of the Aboriginal Sport Circle. Sport Canada, a branch of the Department of Canadian Heritage, was the conduit for this program. The Aboriginal Sport Circle was created in 1992 as a national body to represent the interests of Aboriginal people in sport in Canada.

Federal government support for the North American Indigenous Games (NAIG) commenced with the second North American Indigenous Games in 1993 in Prince Albert, Saskatchewan. Federal government funding continued in 1995, when the NAIG were held in Duluth, Minnesota and again in 1997, when the games were held in Victoria, British Columbia. The federal government contributed in excess of two million dollars to the NAIG held in Winnipeg, Manitoba, in 2002.

The federal government has also supported coaching initiatives for Aboriginal peoples. In 1994, the federal government commenced funding the National Aboriginal Coaching and Leadership Program. The three-year pilot project terminated in March 1997. The program consisted of four elements: an Aboriginal coaching manual, a national Aboriginal Coaching School, Aboriginal coaching clinics, and a National Aboriginal Awards Program.

In additional to federal government programs and initiatives, initiatives and programs for Aboriginal people exist in the territories and in some provinces. The provincial government of Manitoba provides financial support to the Manitoba Aboriginal Sport and Recreation Council. Funding is provided to support the Winnipeg Aboriginal Sport Achievement Center, the Manitoba Indigenous Games, and Team Manitoba (the Manitoba contingent participating in the NAIG).

While there are currently a number of government initiatives in place for Aboriginal sport in Canada, jurisdictional tension exists concerning which level of government should be responsible for funding Aboriginal sport programs and initiatives. The provinces have generally viewed Aboriginal-related issues as falling within the constitutional responsibility of the federal government. The federal government has typically tended to view issues involving sport and Aboriginal peoples as recreational in nature and therefore within the constitutional domain of the provinces. In addition, the Métis, who are included under the umbrella of Aboriginal people, are themselves the subject of a jurisdictional tug-of-war, with the result that Métis athletes are faced with additional funding challenges.

Susan Haslip

See also: Aboriginal Sport Circle; Arctic Winter Games; North American Indigenous Games.

FURTHER READING

Haslip, Susan. "A Treaty Right to Sport?" 8, no. 2 (June 2001) (www.murdoch.edu.au/elaw/issues/v8n2/haslip82.html).

Minister's Task Force on Federal Sport Policy. *Sport: The Way Ahead.* Ottawa: Minister of Supply and Services Canada, 1992.

Paraschak, Vicky. "Knowing Ourselves through the Other: Indigenous Peoples in Sport in Canada." In *Sociology of Sport: Theory and Practice*, edited by R. Jones and K. Armour. Essex: Longman, 2000.

———. "The Native Sport and Recreation Program, 1972–1981: Patterns of Resistance, Patterns of Reproduction." *Canadian Journal of History of Sport 1* 26, no. 2 (December 1995).

———. "Native Sports History: Pitfalls and Promise." *Canadian Journal of History of Sport 57* 20, no. 1 (1989).

GOVERNMENT PROGRAMS AND INITIATIVES—UNITED STATES

U.S. government programs and initiatives introduced American Indians to modern sports. Indians who attended boarding schools learned to play football, baseball, and basketball, as well as a host of other sports, as part of the government's attempt to assimilate Native Americans into the American melting pot. While government officials hoped to transform Indians through sports, Indian athletes imbued sports and these initiatives with their own meaning.

After the Plains Indian wars ended in the late nineteenth century, government officials and individuals sympathetic to the condition of Native Americans cast about for ways to help North America's indigenous population. They settled upon a two-pronged attack on Native American's way of life in an effort to assimilate Indians into white society. The first element, the Dawes Act, allotted reservations and sought to undermine Indian communities. The second, the creation of a nationwide network of Indian boarding schools, taught Indians how to enter American society and culture. Indian boarding schools were the brainchild of Richard Henry Pratt, a former army officer. He and other friends of the Indian believed that the reservation environment was destructive to Indians. They proposed removing Indian children from reservations and placing them in off-reservation boarding schools. There, Indians would be taught how to acclimate themselves to American society. The U.S. government gave Pratt abandoned army barracks in Carlisle, Pennsylvania, and there he started the first Indian boarding school, the Carlisle Indian Industrial School.

In 1893, a group of Indian students approached Pratt and requested that the school sponsor a football team. Pratt recognized the usefulness of athletic contests—they could be turned into another method of assimilating Indians. Pratt made two demands on his Indian football players. First, they needed to compete with and defeat the best teams in the country. By defeating the Yales and Harvards of the nation, Native Americans would demonstrate that they could compete in a modern, industrial society. Second, Pratt demanded that his team demonstrate exemplary sportsmanship and refrain from punching opposing players. Pratt reasoned that if Indians fought on the gridiron, it would show white spectators that Indians had not advanced beyond savagery. Other schools followed Carlisle's lead and Pratt's arguments. Soon, football and other sports programs emerged at the Phoenix Indian School, the Haskell Institute, and the Sherman Indian Industrial School, to name a few.

As sports programs spread across the nation, school officials expanded the scope and usefulness of sports and athletics. Progressives considered outside exercise healthy and attempted to reverse the public health problems endemic to boarding schools. School officials also thought that sports provided an opportunity to inculcate proper gender roles among boys and girls. By the 1930s, big-time sports at the schools had given way to intramural sports. School officials promoted boxing so that Indian boys could demonstrate their masculinity. However, they used the intramural activities as a form of social control for Indian girls. Officials allowed some intramural activities in an effort to control girls' leisure time and prevent sexual promiscuity.

However, Indians supported big-time and intramural sports for reasons that differed from the friends of the Indian. Indians used the athletic experience to enhance their ethnic identity. After Carlisle fired Glenn "Pop" Warner, former players at Carlisle demanded that the school hire an Indian coach. Bemus Pierce, Frank Hudson, and Frank Cayou lobbied to replace Warner. The 1926 Haskell homecoming also demonstrated the intersection of ethnic identity and football. That year, Haskell Institute dedicated a ten-thousand-seat stadium paid for by Native American fundraising and invited Indians from all over the country to attend the game. Native Americans staged a powwow, much to the chagrin of boarding school officials. Some Indian football players were proud of their victories over all-white teams. Indian players, as well as journalists, believed that they were re-creating the Plains wars on the football field. In many of these new skirmishes Indians were coming out ahead. Intramural sports promoted identity formation and Native American pride. For male students, boxing became an important sporting endeavor. Indian pugilists enjoyed the travel, felt pride in their Indian identity, and understood the cultural diversity of Indian country. Sports provided Indians with confidence and stressed pan-Indian connections.

As with most Indian policy measures, the makers of policy and its recipients did not agree on the meaning of sports at Indian boarding schools. Schools officials believed that sports would help assimilate Native American children into mainstream American culture. However, Native American athletes found ethnic pride and opportunities to re-create their Indian identities on the football field and in the boxing ring.

William J. Bauer, Jr.

See also: Assimilation; Boarding Schools.

FURTHER READING

Adams, David Wallace. *Education for Extinction: American Indians and the Boarding School Experience.* Lawrence: University of Kansas Press, 1995.

———. "More than a Game: The Carlisle Indians Take to the Gridiron, 1893–1917." *Western Historical Quarterly* 32 (Spring 2001): 25–54.

Bloom, John. *To Show What an Indian Can Do: Sports at Native American Boarding Schools.* Minneapolis: University of Minnesota Press, 2000.

Oxendine, Joseph. *American Indian Sports Heritage.* Champaign, IL: Human Kinetics, 1988.

Guy Wilder GREEN

Born c.1871
Died Unknown
Baseball team owner

Guy Green was raised in rural Polk County, Nebraska, near the town of Stromsburg. As the founding owner of the Nebraska Indians exhibition team in 1897, Green recognized the potential of a pan-Indian baseball team to tour nationally, drawing large crowds wherever it played. Green developed the Nebraska Indians into one of the most successful of the Indian barnstorming teams, along with John Olson's Cherokee All-Stars of Watervliet, Michigan (1904–1912). For twenty-one years, from 1897 to 1917, the Nebraska Indians averaged over 150 games per season in almost as many towns, often drawing crowds in the thousands, throughout the Midwest, Southeast, and on the East Coast. Green billed his team as "The Only Ones on Earth" and "the Greatest Aggregation of its Kind," thus combining Wild West showmanship with his players' considerable baseball talent. Fortunately, Green also documented the history of the Nebraska Indians in two dime-pamphlets.

Growing up among the Swedish immigrants of Stromsburg, Nebraska, Green developed his love of baseball as a small boy. He played first base on his town team, then for Doane College, in Crete, Nebraska. He received his B.S. from Doane in 1891, played outfield briefly for the University of Iowa, and then returned to his hometown to work at the post office and play amateur ball. Taking his law degree from the University of Nebraska in 1897, he organized the Nebraska Indians hurriedly in June, just after his graduation. Green had noticed that the games at nearby Genoa Indian School were well attended by non-Indians, and he soon realized that he could recruit baseball players from Genoa, Haskell Institute, Flandreau Indian School, and from reservations throughout the Midwest. Green traveled with the team through 1907, recruited, coached, and managed players, kept the books, recorded the game scores and notable events in the team's travels, and profited from the sale of Nebraska Indians pamphlets and postcards, as well as from gate receipts. Soon Green's name appeared on the team postcards as "Sole Owner and Manager" of the team, though he worked during the off seasons as an attorney in downtown Lincoln.

From 1897 to at least 1914 (the last season for which a record is available), the Nebraska Indians established an impressive reputation as one of the most formidable exhibition teams in the country. The cumulative total for these years is a remarkable 1237–336–11, for a redoubtable .786 winning percentage. While the Indians generally avoided embarrassing small-town teams, they trounced Fort Madison, Iowa, in 1898 by the score of 40–4, and Mystic, Iowa, in 1905 by the score of 34–0. A canny scout of baseball talent, Green signed a series of tremendous Indian ath-

Green's Nebraska Indians were one of several barnstorming baseball teams during the early-twentieth century that featured Native American players. *(Nebraska State Historical Society)*

letes, including George Howard Johnson, John Bull Williams, and Jacob Burkhardt.

As the Nebraska Indians succeeded, Green attempted to repeat the success of his novelty team by founding an exhibition team of Japanese ballplayers in 1906. The team quickly folded, but Green then attempted an even greater challenge, purchasing the Lincoln Western Association Club in the fall of 1907. Green's Lincoln team played mediocre ball, and he soon decided to sell the team. In July of 1909, in the middle of his second season, Green completed the sale of the Lincoln franchise to Don C. Despain and Lowell Stoner. Following his marriage to Minnie A. Ericson in 1910, Green gave up traveling with the Nebraska Indians and sold the team in late 1911 or early 1912 to Oran and James Beltzer. In the early 1920s, Green moved his family to Kansas City, Missouri.

Beyond his formation of the Nebraska Indians, Green's greatest contribution to Native American baseball was his recording of anecdotes and the playing history of the team in *The Nebraska Indians: A Complete History* and *Fun and Frolick with an Indian Ball Team*. Without these books, which went through multiple editions and sold thousands of copies, the documentation of Native American barnstorming teams at the turn of the twentieth century would be severely curtailed. While Green recounted the team's experiences with a tendency toward humorous exaggeration and anti-Indian stereotyping, he did accurately detail the talents of his players and the difficult playing conditions they faced on small-town diamonds throughout the United States.

Jeffrey Powers-Beck

See also: Baseball; John Olson.

FURTHER READING
Green, Guy W. *On the Diamond: The Nebraska Indians.* Lincoln: Woodruff-Collins, 1903.

Joseph Napoleon "Joe" GUYON

Born 1892
Died 1971, Louisville, Kentucky
Football player, baseball player

One of the early stars of professional football, Joe Guyon was a multitalented and rugged competitor who could punt, run, tackle, and pass with equal facility.

Guyon, a full-blooded Ojibwe Indian, was born on the White Earth Indian Reservation, Minnesota. His birth name was O-Gee-Chidea, which means "brave man." His father was Charles M. Guyon, and his mother was Mary (maiden name unknown). At 190 pounds and five feet, eleven inches, Guyon was large for his era. He starred as tackle on the famed 1912 Carlisle team, which was led by halfback Jim Thorpe and coached by Glenn "Pop" Warner. It went 12–1–1, averaging thirty-six points a game against top college competition.

In 1913, after Thorpe had graduated, Guyon moved to halfback and helped lead the Indians to a 10–1–1 record. He was selected as a halfback first team on a couple of minor All-American teams and was selected second team on Walter Camp's and Frank Menke's All-American teams.

In the summer of 1914, Guyon left Carlisle to attend Keewatin Academy, in Prairie Du Chien, Wisconsin. In 1917 Guyon entered Georgia Tech and played on Coach John W. Heisman's national champion team, the first southern team to be so recognized. Georgia Tech in that championship season had a 9–0–0 record, averaged 54.6 points a game, and outscored its opponents 491–17. Guyon was named to Walter Camp's "Stars of 1917" squad and was chosen first-team All-American by the *Atlanta Constitution.* Guyon played on the 1918 Georgia Tech team and led the freshmen-dominated team to a war-shortened 6–1–0 record, its only loss being to national champion Pittsburgh. Guyon was named to Frank Menke's All-American team as a first-team tackle.

Guyon joined the emerging professional game in 1919, playing halfback with teammate Jim Thorpe for the Canton Bulldogs for two seasons. In 1920 the Bulldogs joined the newly formed American Professional Football Association (which became the National Football League the following year). He was named to the second-team All-Pro team. In the 1920 season Guyon made a ninety-five-yard punt return, which remained an NFL record for some fifty years. Guyon followed Thorpe to the Cleveland Indians (1921) and then to the all-Native American Oorang Indians based in Marion, Ohio (1922–1923). The Oorang Indians performed to white audience expectations, dressing in Indian regalia and yelling war hoops. Guyon again followed Thorpe to the Rock Island Independents (1924) and then, without Thorpe, played for the Kansas City Cowboys (1925). In 1926 Guyon did not play professional football. His last season of pro ball was with the champion New York Giants (1927).

Near the end of his football career, Guyon was heavily engaged in baseball, most notably playing top-level minor league ball for the Louisville Colonels of the American Association during 1925–1926. The Colonels won the pennant both years, and right

Joe Guyon (left) autographing a souvenir football, poses with fellow hall of famer John Blood McNally following his inauguration in 1966. *(AP/Wide World Photos)*

fielder Guyon's contribution was considerable. In 1925, he played 157 games, hit for a .363 average, 228 hits (thirty-eight doubles, seventeen triples), and stole eighteen bases. The following year, he played 154 games, hit for a .343 average, 209 hits (thirty-six doubles, thirteen triples), and stole twenty-one bases. An injury received while playing baseball ended his athletic career.

Guyon was named to the Pro Football Hall of Fame in 1966, to the National Football Foundation's College Football Hall of Fame in 1971, and the American Indian Athletic Hall of Fame in 1972. Guyon died in Louisville, Kentucky.

Robert Pruter

FURTHER READING

Cope, Myron. *The Game That Was: The Early Days of Pro Football*. New York: World, 1970.

Whalen, James D. "Guyon, Joseph N. 'Joe.' " In *Biographical Dictionary of American Sports: Football*, edited by David L. Porter. Westport, CT: Greenwood Press, 1987.

HASKELL INSTITUTE

The Haskell Institute, a federal Indian boarding school in Lawrence, Kansas, achieved national recognition early in the twentieth century for its highly successful athletics program. Haskell enjoyed its greatest accomplishments in the 1920s, with regionally and nationally competitive football and track teams. Native Americans across the country took pride in the school's successful sports program and demonstrated their support by funding a state-of-the-art stadium, which opened on the campus in 1926.

Like other Indian boarding schools founded in the late nineteenth century, Haskell attempted to assimilate and acculturate Native American children into the predominant Euro-American way of life. Indian boarding schools removed young boys and girls from their families at an early age, immersed them in American "civilization," and suppressed the traditional "savage" Native American culture.

Haskell opened its doors in 1884 and featured only elementary education during its first ten years of existence. The school added some high school classes in the mid-1890s, however, and soon followed with a "normal school," or teachers' education, program. The arrival of older students on the Haskell campus coincided with an explosion of school athletics across the country as high schools, colleges, and universities began to sponsor sports teams for interscholastic competition.

Following in the footsteps of Pennsylvania's Carlisle Indian school, which emerged as a college-football powerhouse at the turn of the twentieth century, administrators at Haskell encouraged the formation of athletic teams. Arguing that participation in popular sports helped to imbue "American" values in Indian students and that successful teams would bring much-needed publicity and public support to Haskell, long-time superintendent H.B. Peairs was among the earliest and strongest proponents of the school's athletic program.

Baseball was one of the first sports to appear in the late 1890s; the institute frequently played local professional minor-league teams. Due to the large number of younger students at Haskell, however, the sports program remained small and local during the first two decades of the 1900s.

Arrival of Big-Time Sports

In 1917, the school discontinued its primary and elementary grades and established a full four-year high school curriculum, as well as a two-year collegiate program. This change in focus ushered in a new era in Haskell athletics. With the school concentrating on older pupils, there were enough male students to field a full complement of athletic teams. Like most other schools of this time period, however, Haskell offered organized sports teams only for men. The school allowed female students to participate in intramural and informal games but barred them from interscholastic competition.

Haskell's male students garnered great publicity and recognition for the school throughout the 1920s. Track and football were usually the most successful sports. The Haskell track team was routinely among the leaders at important track meets like the Drake Relays and the Penn Relays, and it boasted national AAU champions like Philip Osif, All-Americans like Theodore Roebuck, and Olympians like Wilson Charles.

A photo of the 1907 Haskell Basketball team foreshadows the school's later athletic dominance. *(Courtesy of Chicago Historical Society)*

Haskell's student-athletes also achieved success on the gridiron, playing such noted college football teams as Notre Dame, Kansas, Nebraska, Army, Navy, and Michigan State. Throughout the 1920s, the institute fielded competitive squads, and the 1926 team went undefeated, at 12–0–1.

Numerous individual players garnered national accolades for their play on the Haskell football squad. John Levi, Louis "Rabbit" Weller, Albert Hawley, Walter Johnson, and Elijah Smith attracted national recognition during the heyday of Haskell's football program.

The Great Homecoming

American Indians from tribes around the Midwest and West attended Haskell Institute, and Native Americans throughout the country took pride in the accomplishments of Haskell's sports teams in the 1920s. No event symbolized Indians' attachment to the school better than the great Homecoming celebration of 1926. Stimulated by the success of the school's football team, Native Americans sponsored a fund-raising drive in the mid-1920s to construct a new stadium on Haskell's campus. Contributions from tribes and from individual Native Americans raised more than $180,000 for the project.

Haskell inaugurated its new 10,500-seat stadium on October 30, 1926, with a football game against Bucknell University, which Haskell won 36–0. Representatives from nearly eighty tribes came for the festivities, which were highlighted by a grand intertribal powwow.

The cooperation of Native Americans from different ethnic backgrounds in the building of the Haskell stadium epitomized the Indians' changing way of life. Despite the overwhelming success of the fund-raising drive and dedication of the new stadium, however, some federal officials were chagrined by the celebrations, which featured traditional dress, ceremonial dances, and Native American regalia. To these disappointed observers, the Indian boarding schools had been supposed to teach modern American culture to the Indian youth and not to celebrate "savage" traditions.

Decline of Athletics

The opening of the new stadium marked the beginning of the end of big-time sports at Haskell. The Great Depression fundamentally altered the nature of the institute's student body. College-age men left in the early 1930s to find jobs, and many Native American parents sent their younger children to the school to escape the poverty-stricken reservations. With a dearth of older students, Haskell's teams were unable to remain competitive, and the school dropped its collegiate curriculum altogether in 1938.

In the 1940s, after Haskell reverted to a high school, its students continued to find athletic success. The institute's football teams achieved local and state recognition, but it was the school's individual athletes who received the most attention. Nelson Levering, of the formidable boxing program, was a Golden Gloves champion, and runner Billy Mills led the school to a cross-country state championship before earning collegiate and Olympic awards.

In the 1970s, Haskell reinstituted a junior-college program and later changed its name to the Haskell Indian Nations University, once again offering college degrees. In 1999, a new chapter in the school's athletic history dawned when it switched from the National Junior College Athletic Association (NJCAA) to the National Association of Intercollegiate Athletics (NAIA), an organization of four-year schools.

Gregory Bond

See also: Assimilation; Boarding Schools; Carlisle Indian Industrial School.

FURTHER READING

Bloom, John. *To Show What an Indian Can Do.* Minneapolis: University of Minnesota Press, 2000.
Child, Brenda J. *Boarding School Seasons.* Lincoln: University of Nebraska Press, 1988.
Haskell Indian Nations University (www.haskell.edu).
Oxendine, Joseph B. *American Indian Sports Heritage.* Champaign, IL: Human Kinetics Books, 1988.

Pete HAUSER

Born c.1884, El Reno, Oklahoma
Died July 21, 1935, Osage County, Oklahoma
Football player

Of German, Cheyenne, and Arapaho heritage, Hauser was one of five children born to army sergeant Herman Hauser and Amy Broken Cup. With his brother Emil—a prominent athlete in his own right—Hauser attended Haskell Institute in Lawrence, Kansas, from 1901 to 1906 and the Carlisle Indian School in Pennsylvania from 1906 to 1910. He played football and baseball at both institutions, earning varsity letters for

Pete Hauser (with the ball) was described by legendary coach Pop Warner as the greatest fullback to play at Carlisle Indian Industrial School. *(Courtesy of Chicago Historical Society)*

the former as a member of the Haskell Indians (1904 and 1905) and the Carlisle Red Men (1906–1910). At Carlisle, Hauser played fullback; he was captain of the varsity team in 1910 and was granted honorable mention as an All-American athlete that same year. His coach at Carlisle, Glenn S. ("Pop") Warner later described Hauser to a *Carlisle Herald* reporter as a great fullback—probably the best the school had ever known; he credited Hauser with having thrown the first spiral pass and compared him favorably to Jim Thorpe in all sports except for track. Hauser was eventually named to the All-Time Greatest Team of Carlisle Red Men.

The Carlisle team of Hauser's era was among that institution's best, and it defeated several of the best college teams in the country. Even so, it suffered discrimination. On numerous occasions, players responded with pointed historical and cultural observations. According to Glenn Warner, in the midst of a particularly ferocious game, when Hauser was unfairly kneed by a member of the opposing team, he responded by quipping, "Who's the savage now?"

After leaving Carlisle, Hauser played professional football in Detroit for several years before joining the American Expeditionary Force during World War I. On his return to Oklahoma at the close of that conflict, Hauser organized several semiprofessional Indian football teams in and around Osage County and was employed by the Indian Emergency Conservation Corps. He died on July 21, 1935.

He was inducted into the American Indian Hall of Fame in 1972.

Michael Sherfy

FURTHER READING

Adams, David Wallace. "More than a Game: The Carlisle Indians Take to the Gridiron, 1893–1917." *Western Historical Quarterly* 32, no. 1 (2001): 1–32.

Hauser, Pete. "Christians at Home." *Carlisle Arrow*, December 31, 1909.

Hauser, Pete, and Joseph Libby. "Practical Business Education." *Carlisle Arrow,* April 8, 1910.

Albert M. HAWLEY

Born 1906, Hays, Montana
Football player

Hawley (Gros Ventre-Assiniboin) was a collegiate football star and later official with the Indian Service.

Hawley attended Haskell Institute between 1920 and 1928. Known as the "Montana Bull" by his teammates and fans, he first played on Haskell's reserve football team in 1924 before joining the varsity squad as center a year later. The *Haskell Annual* praised Hawley's leadership abilities: "Almost immediately he won the respect and confidence of his new teammates and was an inspiration to them throughout the season." At the end of the 1927 season, national college coaches invited Hawley to play in the annual East-West game, held in San Francisco, California.

In 1928, Hawley transferred to Davis and Elkins College in Pennsylvania, where he also played football. He won All-American honorable mention honors in two seasons at Davis and Elkins.

After his playing days, Hawley went to work in the Bureau of Indian Affairs. He was a teacher, coach, and principal in Indian schools in Idaho and Nevada. He also served as a reservation superintendent in Arizona. He also held various positions outside the Indian service, including serving as the American Athletic Union's boxing commissioner in Nevada. In the 1970s, Hawley was appointed to the electoral board of the American Indian Athletic Hall of Fame, and in 1973 he himself was inducted to the Hall of Fame. In 1978, he was elected to the Davis and Elkins sports Hall of Fame.

William J. Bauer, Jr.

FURTHER READING

Machamer, Gene. *The Illustrated Native American Profiles.* Mechanicsburg, PA: Carlisle, 1996.

Oxendine, Joseph. *Native American Sports Heritage.* Champaign, IL: Human Kinetics, 1988.

"World of Sports Enriched by 61 American Indian Athletes." *Journal of American Indian Education* 10 (May 1971).

HERITAGE

If one considers sports to be competitive organized activity, Native heritage and sports may seem to have little to do with one another. Traditionally, the physical activities of skill we would today make competitive and call sports were skills needed to survive, which were often believed to come with the assistance of the spirit world. Consider, for example, the tale of how the bow and arrow came to be. A man was out hiking in the woods and came upon a bear. Because he did not want to fight the bear, he attempted to run away. However, the bear, knowing he was stronger than the man, chased him. As the man ran through the bushes his spear got caught on a vine; he had to use all of his strength to try to free it. Just as the bear caught him the hunter lost his grip, and the vine, becoming a natural sling, threw the spear into the bear's chest and killed it. The hunter thanked the Great Spirit for saving his life and creating a new weapon. Children were then schooled in how to use this weapon, which became a vital part of life. Today most people do not use a bow and arrow for protection or to get food but might use one to shoot at a target or com-

One-foot high kick, a traditional contest and current event at the World Eskimo-Indian Olympics, is a good example of the lasting importance of indigenous sporting heritage. *(Paul A. Souders/CORBIS)*

pete with one another. In short, what once was a skill required for survival we consider today a sport.

This being the case, perhaps it is not surprising that the Native idea of sport and the ideas of sports brought over by Europeans as they colonized the New World are quite different and often even seem to clash. Members of a Native community often needed to rely on one another for survival, as they all hunted and gathered together; they developed a culture prohibiting the idea of winning, of one being the best. For this reason sporting competitions were generally a matter of timing; success could be as much the result of the loser's exhaustion as the winner's superior skills. For example, a wrestling competition would start with the smaller boys in the group. When one boy was thrown a third would enter the ring. If the third boy threw the first, he would remain in the ring and take on all comers until he was himself thrown. Otherwise the first boy would stay until someone could throw him, as a result of superior size or skill or because the original competitor was just worn out. In either case the competition would continue in this fashion, with the participants getting progressively bigger and older until only one man was left. Thus, the competition had no junior or senior categories, and the only real organization was that everyone was to compete with someone about his own size, which kept the competition fair.

This lack of categories and seeming lack of organization has generated problems when Native culture intersected with modern sporting competitions, with its tournaments, categories, and rigid timetables. For while the cultures inspired by Europe believe in the importance of competition to find and recognize the best, sporting tournaments inspired by Native cultures often see sports as part of broader cultural festivals to celebrate their heritage, with powwows, singers, prayers, and storytelling as well as demonstrations of sporting skill. It is a time to show any and all talents. However, more often than not, governments or other organizations that fund such events see the former, European view as correct and often show little patience with the latter.

Nevertheless, it must be remembered that indigenous individuals of today are not simply influenced by their traditional cultures but are exposed to, and often have imposed on them, society's culture. This is as true with sports as with anything else. Perhaps the most ironic example of this is the Good Woman contest, which was put in place to replace the selection process for the Indian princess. In this contest women would participate in a variety of activities that represented the skills a "good woman" would need to use in her daily endeavors. As time passed categories were imposed on this event. These categories were not just age based but opened the competition to males. This was acceptable to those representing the societal culture who funded the events; however, to indigenous individuals male participation was more a subject of laughter than something to be judged seriously. In short, the traditional choosing of an Indian princess was changed to a competition and then transformed by categories that allowed men to compete—and all of this was the result of requirements placed on the event by nonindigenous influences. Yet it would be an error to believe that these outside influences meet no resistance, even as they operate to make indigenous games more similar to games following the European tradition. Sometimes this resistance may be as simple as laughter, as in the case above, but it can be more direct, such as the ignoring of time schedules by participants who sleep in after a late night of cultural festivities (such as dancing). Put simply, if no one shows up in the morning because they are all sleeping, there can be no athletic competition—often to the extreme frustration of nonindigenous organizers.

Despite the fact that nonindigenous cultures have altered their nature, these events remain vital to the well-being of indigenous communities and their youth. Sports have proven to be a good way to combat problems plaguing reservation communities. What is more, sporting activities combined with broader cultural festivities can assist the young in connecting with their heritage.

Kathy Collins

See also: Adaptation; All Indian Competitions; Assimilation; Competition Powwows; Gender Relations; Iroquois Nationals Lacrosse.

FURTHER READING

Oxendine, Joseph B. *American Indian Sport Heritage.* 2d ed. Lincoln: University of Nebraska Press, 1995.

Paraschak, Victoria. "Variations in Race Relations: Sporting Events for Native Peoples in Canada." *Sociology of Sport Journal* 14 (1997): 1–21.

HOCKEY

The modern sport of ice hockey is derived from a variety of games played in both North America and Europe, including contests conducted by Native Americans and Canadians. The Micmac tribe of Nova Scotia, among others, played a stick and "puck" game on ice as early as the 1600s.

According to conflicting and sources, the name "hockey" may have come from either the Mohawks of the northeastern United States or the Iroquois of the St. Lawrence River Valley. Stories of the tribes calling the game *hoghee* (loosely, "it hurts"), which evolved to "hockey," may well be apocryphal, but archeological evidence shows that both tribes played hockey-like games. Other reports ascribe the name to a French game played with a curved shepherd's crook called a *hoquet.*

Organized hockey leagues and codified rules developed by the turn of the twentieth century. Native Americans and Canadians were often openly or informally excluded from leagues organized by those of European descent. Meanwhile, some teams, including the Chicago Black Hawks of the National Hockey League, adopted "Indian"-derived names or emblems.

It is probable that some lighter-skinned

players of mixed heritage competed in prominent hockey leagues without acknowledging their roots. There were also teams formed in Native communities. All-Aboriginal teams and tournaments exist to this day. The first player from a Native Canadian community to reach the National Hockey League was Fred Sasakamoose of Debden, Saskatchewan. The player cracked the league for a brief, eleven-game, run with Chicago in 1953–1954.

Sasakamoose, like subsequent players of aboriginal descent, dealt with both mean-spirited and good-natured racism on and off the ice. A large percentage of players of full-blooded or mixed Native heritage have been nicknamed "Chief," a sobriquet usually intended with no overt malice. Most players given the nickname, including Boston Bruins standout John Bucyk and George Armstrong of the Toronto Maple Leafs, have carried it with pride.

More troubling is the still-prevalent stereotype of the Native player as an unskilled, brawling, "goon" on the ice. While many players of partial or full Native descent, including Stan Jonathan, Craig Berube, Gino Odjick, Denny Lambert, Sandy McCarthy, and Scott Daniels, have been better known as "enforcers" than skill players, the image masks both the widespread popularity and respect attained by these players and the accomplishments of more skilled, finesse-oriented competitors, including Armstrong, Reggie Leach, and Hall of Fame players Bucyk and Bryan Trottier. Chris Simon, meanwhile, combined both toughness and goal-scoring ability. Some, like former players Armstrong and Ted Nolan, have also been successful head coaches.

Many players of Native Canadian descent have taken on active community or political leadership roles. Jonathan campaigns for the recognition of treaty rights, and Nolan is deeply involved in athletic, communal, and educational causes for members of First Nation communities.

Alcoholism, a disease often associated with Native communities, has touched the lives of many Native players, although it should also be noted that alcohol abuse is prevalent in sports in general and hockey in particular. Leach, Simon, and Odjick, among others, have conquered alcohol dependency and taken active roles in speaking to members of First Nation communities and youth in the general population about their experiences.

William R. Meltzer

See also: Ice Shinny.

FURTHER READING

Dolan, Edward. *The National Hockey League.* Greenwich, CT: Bison Books, 1986.

They Call Me Chief. Written and directed by Don Marks. Gary Zubec, producer. Maple Lake Releasing, 1998.

Elon Chester "Chief" HOGSETT

Born November 2, 1903, Brownell, Kansas
Died July 17, 2001, Kansas
Baseball player

Hogsett, a Cherokee pitcher, played professional baseball for eleven seasons (1929–1938, 1944). His lifetime record was 63–87, with a 5.02 ERA.

One of eleven siblings, the youth hated the family farm and his carousing stepfather. He left home at fourteen and never came back, though he pitched for the Brownell high school team and various town ball clubs. One legacy of the farm, though, was Hogsett's underhanded motion, which came from skipping stones in boredom.

The Detroit Tigers picked up Elon's contract in 1925. After he won twenty-two games at Montreal in 1929, the Tigers called him up, and he made his major-league debut on September 18. Over the next few years, Hogsett developed into Detroit's main lefty reliever. He allowed just one run in seven and one-third innings in the 1934 World Series, which the St. Louis Cardinals won, and added another scoreless inning as the Tigers topped the Cubs in the 1935 Series. Detroit dealt Elon to the St. Louis Browns the next year, and in 1937 he endured a miserable year for the impoverished last-place club.

Hogsett pitched from 1939 to 1944 in the American Association with Minneapolis and Indianapolis. The wartime shortage of bodies won him a final call-up with the Tigers in 1944. Elon then called it quits and returned to Kansas, where he lived out the rest of a remarkably long life. Although for many years he was self-sufficient, Alzheimer's eventually forced him to move to a rest home, where he passed away on July 17, 2001.

Rory Costello

FURTHER READING

Auker, Elden, with Tom Keegan. *Sleeper Cars and Flannel Uniforms*. Chicago: Triumph Books, 2001.

Bak, Richard. *Cobb Would Have Caught It: The Golden Age of Baseball in Detroit*. Detroit: Wayne State University Press, 1991.

Kaufman, James C., and Alan S. *The Worst Baseball Pitchers of All Time: Bad Luck, Bad Arms, Bad Teams, and Just Plain Bad*. Jefferson, NC: McFarland, 1993.

HOMINY INDIANS

The Hominy Indians were an all-Indian football team from Oklahoma that played semiprofessional and professional teams with spectacular success in the 1920s and 1930s. The idea of an all-Indian professional football team in Oklahoma was the brainchild of Ira Hamilton, a young Osage living in Hominy in northeastern Oklahoma in the early 1920s. Hamilton bought the first uniforms and conducted a series of practice scrimmages. The first coach of record for the Hominy Indians was Pete Hauser, former Carlisle and Haskell Institute star. He guided the team for two undefeated seasons against town teams in Missouri, Kansas, and Oklahoma.

By 1925, the Hominy Indians were winning so consistently that it was necessary to find new teams to play. A group of oil-rich Osage Indians offered financial backing. Through the influence of these tribal leaders, Hominy was able to field teams strengthened with players graduated from Haskell and other colleges. It was a matter of pride that a college graduate-Indian could "play for pay" with an all-Indian professional football team.

Ex-Haskell All-American John Levi and his brother George joined the team, John as a player-coach. Many tribes were represented on the Hominy roster, including Osage, Pawnee, Otoe, Creek, Seminole, Cheyenne, Arapaho, Sioux, Cherokee, Navajo, Kiowa, Seneca, Sac and Fox, and Pottawatomie. One outstanding player was an Inuit from Alaska.

After running up a string of twenty-eight consecutive victories, Hominy challenged the barnstorming New York Giants, newly crowned champions of the NFL. The game was played the day after Christmas, 1927, in the oil-boom town of Pawhuska, Oklahoma, the capital of the Osage tribe. The Giants scored first, but the Indians struck back when an alert Hominy end grabbed a Giant fumble in midair and raced fifty yards for a touchdown. Deep in the third quarter, Pappio outraced the Giant safety and caught a sixty-yard pass from John Levi for the winning touchdown in a 13–6 Indian victory.

The Hominy Indians did not continue winning forever. In 1928, a hand-picked team of NFL All-Stars defeated the Indians in Tulsa, Oklahoma, 27–0. In 1931, the New York Giants got their revenge by defeating the Indians 53–0 at the Polo Grounds of New York City. From 1929 though 1932, the Indians traveled from coast to coast. A few games were played in California, but the majority were played in the East. There were games in St. Louis, Boston, New York City, Chicago, and many smaller cities. The Indian team from Oklahoma had little trouble scheduling games, since its members were splendid, colorful athletes.

With the advent of the thirties, the Great Depression spread across the land. The golden age of pre-Depression sports was at an end. The Hominy Indians disbanded after the 1932 season.

Royse Parr

See also: Football; Oorang Indians.

FURTHER READING

Shoemaker, Arthur. "Hominy Indians." *Oklahoma Today* 17, no. 4 (Fall 1967).

HOOP AND POLE

Hoop and pole was a game played throughout North America but primarily contested by the tribes in what is now the United States. Many variations of the game have been reported, but the main point of the game is consistent throughout the continent. The object of the game was to throw a long spear, dart, or arrow at a rolling hoop. The game was played strictly by men. All the tribes that played this game attached great religious and cultural significance to it.

The pole or darts varied in length from tribe to tribe from as little as two feet to as much as twelve feet; some tribe used even longer poles. Poles were ornately decorated with etchings, along with feathers, buckskin strings, or the claws of animals. Many poles were affixed with barbs designed to catch in the netting within the hoop. In some variations of the game the players used shorter dartlike projectiles made of corn cobs and feathers. Some tribes simply threw or shot arrows at the rolling hoop. The Apache and Navajo used long jointed poles.

The hoop was usually made of a sapling bent into a circular shape between six and twelve inches across and bound with a mesh of rawhide. The rawhide was woven into a web within the bent circular sapling. The hoop could also be twined with a cord, beads, or cornhusks. A few tribes made their hoops out of stone. The webbing was woven into various patterns. Other hoops were divided into simple quarters or halves. Some tribes attached beads of different colors at equidistant points along the hoop.

The game was played on a flat large field with as many stones or pebbles removed from the playing area as possible. Playing fields varied from tribe to tribe; for instance, the Mandan of North Dakota were said to play on timber floors 150 feet long, while the Apache court had a rock in the center of the field from which the players threw the poles at the hoop. The Creeks had enclosed courts with upward-sloping sides that served as bleachers for spectators to view the action. Some tribes lined the field with clay or sand or some other additive to ensure that the hoop rolled as smoothly as possible. The aim of the game was to hit the rolling hoop with the pole, preferably in the middle of the hoop. Some tribes awarded higher point totals for putting the pole in particular holes.

Usually, the game was played by teams

of two to four players, with two being the most common number of participants. In most cases the teams stood side by side, throwing their projectiles at the rolling hoops. A few tribes stood in two parallel lines shooting at the hoop.

The Apaches and many other tribes used the game for gambling purposes. Most importantly, though, the game had great social, cultural, and religious significance across the continent. The hoop and its webbing signified different things to different tribes. The Zuni hoop and webbing was said to represent the shield of the twin war gods Ahaiyuta and Matasailema as woven by their mother the Spider Woman. The Navajos believe that the game was given to them by the ancient spider people.

Hoop and pole was a popular game that Native Americans played for its sheer enjoyment. However, the strong religious overtone of hoop and pole was of vital importance to the players and to the cultures of the tribes that played the game.

Rick Dyson

FURTHER READING

Culin, Stewart. *Games of the North American Indians.* New York: Dover, 1975.

Oxendine, Joseph B. *American Indian Sports Heritage.* Champaign, IL: Human Kinetics, 1988.

Waneek HORN-MILLER

Born November 31, 1975, Kahnawake, Quebec
Water polo player

Waneek Horn-Miller, a Mohawk, was born in Kahnawake, Quebec, as one of four daughters of longtime Native activist Kahn-Tineta Horn. Always keenly athletic, she began swimming competitively at age seven, eventually winning some twenty gold medals at the North American Indigenous Games. In 1989 she was an Ontario age-group champion; that same year she began playing water polo on her high school team in Hull, Quebec. In 1993, she made the national junior team, helping it finish fifth at the Junior World Championships that year in Quebec City. After graduating from high school, she attended Carleton University in Ottawa; there she played on the women's water polo team and was named Female Athlete of the Year three years in a row. Promoted to the senior national team in 1995, she helped Canada win a gold medal at the 1999 Pan American Games by scoring three goals in an 8–6 win over the United States.

The team earned a berth into the 2000 Summer Olympics in Sydney, where women's water polo was on the program for the first time. With Waneek as co-captain, the team came in fifth out of six nations. As the first Canadian woman of Mohawk heritage to compete in the Olympic Games, she proudly adorned the cover of *Time* magazine—naked except for a water-polo ball concealing her breasts (*Time*, Canadian Edition, September 11, 2000).

Much of her drive and determination to succeed stems from her experience as a teenager. In the summer of 1990, when she was fourteen, she accompanied her mother to the Mohawk community of Kanesatake, also known as Oka, about thirty miles north of Montreal. The lands allocated to the Mohawks do not officially constitute a reserve and are interwoven with lands belonging to nonaboriginal people of the village and parish of Oka. The "Oka Crisis" erupted when the Mohawks tried to stop a golf course from encroaching on their ancestral burial grounds at Oka; a bloody confrontation ensured between natives, Quebec's provincial police, and the Canadian army. As she tried to leave the scene

Waneek Horn-Miller celebrates with teammates their gold medal victory at the 1999 Pan Am Games. *(AP/Wide World Photos)*

with her four-year-old sister, Horn-Miller was stabbed in the chest by a soldier's bayonet, leaving physical and emotional scars. "I grew up during the crisis," she later observed. "I just realized how important being Native was and how important our culture is." In the years that followed she overcame her anger, confusion, and bitterness with support from family and friends, and by focusing on school and athletics. "I could've become really racist and done nothing with that experience," but as she explained: "I could've really isolated myself, but went through Oka for a reason and I decided not to let it hinder me; I'm going to let it do something for me."

Today, Waneek still trains hard as a member of the Canadian national women's water polo team. She graduated from university with a degree in political science and works part-time with the Aboriginal Peoples Television Network hosting a program about up-and-coming artists and musicians in the Aboriginal community. As a role model for native youth, she often speaks out about the need to stay in school, work hard, and achieve goals. In 2000, her efforts and accomplishments were recognized by a prestigious National Aboriginal Achievement Award. "Because of Oka, I've learned that life is too precious and too short to waste," Horn-Miller observes. "Look on the positive, see the good in people, see the potential in yourself and others."

M. Ann Hall

FURTHER READING

Starkman, Randy. "From Oka Battles to Pan Am Glory." *Toronto Star*, July 29, 1999, A1.

Stubbs, Dave. "Their Goal Is Gold." *Time* (Canadian Edition), September 11, 2000, 60–61.

Wong, Christine. "Oka Vet Water Polo National." *WindSpeaker* (May 1996).

Stacy HOWELL

Born 1904
Died 1996
Football player

Stacy Howell, a Pawnee, became the first All-State football player selected from Pawnee, Oklahoma, and later had a brief college football career. Although such feats are commonplace, Howell accomplished all this after losing his right arm in a hunting accident.

In the early 1920s, Howell was a promising offensive lineman at Pawnee High School. Following his sophomore season, he lost his right arm while duck hunting with some friends. On bed rest for some time, Howell was at first forlorn, but encouraged by his coach, Mose LeForce, he returned to the team his senior year, wearing a special pair of shoulder pads, one with a heavy leather cup that protected the stump of his right arm. Despite his handicap, Howell was named to the All-State team.

Howell secured a scholarship from Franklin College. Though he performed well on the team, he was often homesick and returned to Pawnee when possible. Howell met his future wife, Rebecca Beatty, while hitchhiking home. Howell and Beatty had four kids and remained married for over sixty-nine years.

Later in life, Howell, who received his college degrees while in his late forties and early fifties, served as an educator (teacher, principal, and superintendent) at the Indian schools in the Santo Domingo, Laguna, and Taos Pueblos and in Albuquerque. He also remained active in sports, playing tennis and golf, and serving as a referee and umpire in local communities. In 1994 the Pawnee Tribal Business Council honored Howell by establishing an academic/athletic award in his name.

M. Todd Fuller

Stacy S. "Bub" HOWELL

Born 1928
Basketball player

Stacy S. Howell, a Pawnee, gained fame as an outstanding collegiate basketball player in the late 1940s and early 1950s, which culminated in his being named to the National Association of Intercollegiate Athletes (NAIA) All-American team in 1950 from East Central College in Ada, Oklahoma. In 1977, Howell was inducted into the American Indian Athletic Hall of Fame.

The oldest of four children (three boys and one girl), Howell excelled at many sports, including football and baseball. After his high school graduation, he attended Murray A & M Junior College in Tishomingo, Oklahoma, where he earned all-conference honors in 1947 and 1948. In 1948, Howell was also selected as a junior college All-American.

Howell then attended the University of Idaho on a basketball scholarship (along with two other teammates from Murray A & M) but, according to his sister, Betty Evans, did not like being so far away from home and transferred to East Central College in Ada, Oklahoma. In 1949, he won All-Conference honors (second team) with East Central College. During his year, in 1950, Howell was named to the Helm's

Foundation All-American team. He was also given All-American honors for his performance at the Oklahoma AAU tournament. After his collegiate playing days, Howell played semipro basketball for a short time (one year) with the Haliburton (Oklahoma) Cement Factory team.

Completing his studies at East Central, Howell became a history teacher and high school basketball coach for different schools in southern Oklahoma. After performing such duties for a few years, he moved to Santa Fe, New Mexico, where he taught and coached at a number of the Pueblo schools. Later, Howell moved to Oxnard, California, where he also taught and coached.

M. Todd Fuller

Frank HUDSON

Born 1875, Paguate, New Mexico
Died Unknown
Football player

Hudson (Laguna Pueblo) attended the Carlisle Indian Industrial School between 1890 and 1900, playing football from 1895 to 1898, and in 1900. During his time at Carlisle, the team posted a 27–20–1 record. Hudson played quarterback and kicker for the Indians.

Hudson practiced dropkicking year round and was an ambidextrous kicker. By 1897, Hudson had established himself as one of the finest dropkickers in the country. A jeweler in Carlisle, Pennsylvania, rewarded Hudson with a gold ring for his dropkicking exploits against Yale that year. At the end of the season, Walter Camp named Hudson to his second team All-American team, and the following year Hudson served as team captain.

Hudson remained at Carlisle after graduating. In 1905, he served as an assistant football coach to George Woodruff. That season, he appeared in Carlisle's game against Massillon Athletic Club but broke his nose on the fourth play of the game. The following year, he was Bemus Pierce's assistant coach. He also pitched for the employee baseball team, which often scrimmaged Carlisle's team. In 1914, he lobbied to succeed Pop Warner as Carlisle's head coach, but school officials did not hire him. He later became a bank clerk in Pittsburgh, Pennsylvania. Hudson was elected to the American Indian Athletic Hall of Farm in 1973.

William J. Bauer, Jr.

FURTHER READING

Adams, David Wallace. "More than a Game: The Carlisle Indians Take to the Gridiron, 1893–1917." *Western Historical Quarterly* 32 (Spring 2001): 25–54.

Newcombe, Jack. *The Best of the Athletic Boys: The White Man's Impact on Jim Thorpe*. Garden City, NY: Doubleday, 1977.

Oxendine, Joseph. *American Indian Sports Heritage*. Champaign, IL: Human Kinetics, 1988.

Steckbeck, John. *Fabulous Redmen: The Carlisle Indians and Their Famous Football Teams*. Harrisburg, PA: J. Horace McFarland, 1951.

Howard "Cactus" HUNTER

Born February 4, 1951, Pine Ridge Reservation, South Dakota
Rodeo rider

Hunter (Lakota) was born on the Pine Ridge reservation in South Dakota. Rodeo was his primary interest from a very young age. He began riding saddle broncs at the age of ten and soon grew into high school rodeos. Hunter competed almost continuously for several years on the amateur rodeo circuit—riding both saddle broncs and bulls—before joining the Professional Rodeo Cowboys Association in 1970.

Hunter broke into professional rodeo with the support and guidance of Shawn Davis, a three-time world champion saddle bronc rider in his own right. In 1971, Hunter quit riding bulls to focus exclusively on saddle broncs. His specialization paid off; Hunter went on to qualify for the National Finals in Oklahoma City three times (1976, 1979, and 1980) and the Indian National Finals in Salt Lake City five times. Hunter was also champion of the South Dakota Rodeo Association several times.

Before being forced into retirement by an injury in 1995, Howard Hunter was awarded over fifty buckles as awards for competing in professional rodeos and twenty-seven feathers for competing in all-Indian events. Hunter now lives in Kyle, South Dakota, with his wife, Betty Ann ("Annie"). Their two children and their families live nearby.

In 2002, Hunter was recognized for his impact on the rodeo world by being inducted into the Cowboy Hall of Fame in Pierre, South Dakota. He was also identified by name that same year in a commemoration of rodeo by South Dakota's legislature. Several of today's young rodeo stars, including champion-rider Tom Reeves, cite Howard Hunter as one of their idols and rank him among the greatest competitors their sport has ever known.

Michael Sherfy

FURTHER READING

Howard Hunter Collection, newspaper clippings. Oglala Lakota College Archives. Kyle, SD.

"Interview with Tom Reeves." *Indian Rodeo News.com* (www.indianrodeonews.com/Tom.htm), November 12, 2002.

Sioux Nation Cowboy News 1, no. 2 (February 1980).

Sioux Nation Cowboy News 1, no. 10 (October 1980).

South Dakota Senate Commemoration 14 (legis.state.sd.us/sessions/2002/bills/SC14enr.htm), 2002.

ICE SHINNY

Ice shinny, often identified as the precursor of ice hockey, was a game played by Native Americans of the northern Plains and Canada. The game was popular among the Sioux, Crow, Blackfoot, and the numerous tribes of Canada. A variation of ice shinny is still played in Canada today. Two teams of a varying number of players, usually numbering between ten and fifty team members each, played on a large rectangular ice surface. Teams were chosen by various methods, but one of the unique methods involved players placing their shinny sticks into a pile. Then a non-playing member of the tribe pulled two sticks from the pile for each team, placing the sticks into two piles, one to their left and one to their right. The sticks in the pile on the left played on one team, while the sticks in the pile to the right made up the opposing team. The game was played on virtually any frozen-water surface, be it a pond, lake, or river. Two upright logs served as goalposts; each team defended its own goal. Each team's goal posts were hammered into the ground ten to twenty feet apart; opposing team's goals were usually a quarter-mile apart. The object of the game was to hit the "puck," a rawhide covered round knot of wood or a spherical stone, into the opponent's goal. The pucks varied in size from the size of a modern

A scene depicting ball play in the heart of winter. *(Western History Collection, University of Oklahoma Library)*

day golf ball to a size slightly larger than a baseball and were often decorated with ornate painting. The team that scored seven goals first was declared the winner.

Players used a curved stick or shinny thirty to thirty-six inches long to hit the puck. The end of the stick used to strike the puck was flattened and widened much like modern hockey sticks. The sticks were personalized by each player, either with bright paint or etchings. The puck could be struck only with the shinny stick or kicked; players were not allowed to touch the puck with their hands.

Although everyone played ice shinny, women participated more than men, competing against other teams of women. It could be played purely for fun or in fierce competition between rival tribes. Single games often lasted for hours, with no break in play. Games between women and children were less physical than men-only games.

Rick Dyson

See also: Shinny.

FURTHER READING

Culin, Stewart. *Games of the North American Indians*. New York: Dover, 1975.
Oxendine, Joseph B. *American Indian Sports Heritage*. Champaign, IL: Human Kinetics, 1988.
Whitney, Alex. *Sports & Games the Indians Gave Us*. New York: David McKay, 1977.

IROQUOIS NATIONALS LACROSSE

The Haudenosaunee, "People of the Longhouse," or Iroquois Confederacy, consisting of the Mohawk, Oneida, Seneca, Tuscarora, Onondaga, and Cayuga groups, played *Guhchigwaha*, or "bumps hips," an ancient Native American sport first seen by French settlers, who named the game lacrosse. As more settlers learned to play, the sport's popularity rose, and it even became the national game of Canada in 1867. In the late nineteenth century society generally disapproved of playing any sport for money. Without a source of income few people could afford to buy equipment, travel, or pay entry fees, especially reservation-bound Native Americans. Native groups were prohibited from international competition in 1880, because they accepted money to help finance trips. In the 1920s the Iroquois played box lacrosse, which was played inside a rink with fewer people, forming a box lacrosse league in the 1970s, but the number of Native peoples participating in lacrosse declined, while the number of wealthy people and prep-school players increased.

In 1983 Iroquois Nation members Oren Lyons (former lacrosse All-American), Wes Patterson, and Rick Hill formed the Iroquois Nationals Lacrosse team to promote the sport, reclaim part of their heritage, and help instill Iroquois youth with a positive self-image and understanding of their history. The modern game of lacrosse plays as an important spiritual and psychological role in the lives of the Iroquois people as the ancient games did; it is an essential part of the Iroquois creation story and considered a gift from the Creator. The sport is also considered a "medicine" game, because playing also helped to build stamina and strength for hunting and fighting; it was also played to cure and prevent illness.

The ban prohibiting the Iroquois people from international play lasted until 1987. When the Iroquois Nationals League was formed in 1983 the founders immediately sought acceptance into the International Lacrosse Federation. They especially wanted to be accepted by 1986 so they would be eligible to compete in the 1986 world championship in Toronto. The federation turned them down in 1986, so the Iroquois hosted an exhibition tournament at the State University of New York in Buffalo to play the Australians and the English before the Toronto tournament. The Iroquois teams played a game of "fireball," a game similar to soccer but played at night with a flaming kerosene-soaked ball, for the English and Australians.

Having proven themselves both competitive and organized enough to participate on the international level, the Iroquois Nationals were admitted to the federation in 1987 and invited to play in the 1990 World Games in Perth, Australia. Moreover, the federation recognized the Iroquois as not only a team but a member nation, making the Iroquois Nationals the only indigenous nation competing in international sport in the world. The United States opposed the Iroquois bid for status as an international team, but when the federation accepted the Iroquois teams Washington reluctantly agreed. Since their first appearance in the world games the Iroquois Nationals have regularly appeared in the top five finishers.

The Iroquois Confederacy is a sovereign nation and does not consider itself part of Canada or the United States, or its people citizens of either country; its reservation is both in northern New York and southeastern Canada. When the team travels it does so with Haudenosaunee passports, which are recognized by a number of foreign governments. They also have their own flag and their own national anthem. A volunteer board of directors, representative of the different nations of the Iroquois Confederacy, runs the Iroquois Nationals. Board members include NCAA lacrosse champions, LAX All-Americans, stick makers, and U.S. and Canadian hall of fame inductees. Community as well as private and corporate sponsors fund the organization. Under their logo, the eagle dancer, the traditional symbol of spiritual strength, the Iroquois Nationals also conduct clinics for Native groups.

Lacrosse has become one of the fastest-growing team sports. Almost four hundred colleges and universities and 1,500 high schools have men's lacrosse teams. Although lacrosse is traditionally a male sport, women also play. In the new century the Iroquois Nationals continue to play an important part of Iroquois culture, history, spirituality, and self-image, linking the Iroquois to their past.

Lisa A. Ennis

See also: Heritage; Lacrosse; Nationalism.

FURTHER READING

Brady, Erik. "Iroquois Reach for the Moon; Indians Carry Flag into Competition." *USA Today,* July 5, 1990.

Hoyt-Goldsmith, Diane. *Lacrosse: The National Game of the Iroquois.* New York: Holiday House, 1998.

Iroquois Nationals Lacrosse. "The Nationals Lacrosse Program" (www.iroquoisnationals.com).

Lipsyte, Robert. "All-American Game." *New York Times,* June 15, 1986, sec. 6, 28.

Jack "Indian Jack" JACOBS

Born 1919
Died 1974
Football player

Jack Jacobs, a Creek, may be the best remembered of many outstanding Indian athletes to have played sports at the University of Oklahoma (OU). In the opening game of his college football career in 1939, he returned the opening kickoff sixty-eight yards against Southern Methodist University in a 7–7 tie game. Some observers opined that Jacobs could do anything he wished on the gridiron. He led OU to victories in the next six games before OU finished the season with losses to Missouri, 7–6, and Nebraska, 13–7.

In his junior year in 1940, Quarterback Jacobs displayed a passing ability the likes of which had never been seen at OU. In the season opener against rival Oklahoma A & M, OU outlasted the Aggies 29–27 on the strength of Jacobs's arm. He played his best game of the year the following week against Texas University, but it was not enough; OU lost, 19–16. During 1940, he established a college punting record of 47.8 yards per kick.

With a new coach and formation in 1941, Jacobs led the OU team to a 6–3 record for the year. As the team's punter, he set a single-game record on a rainy day against Santa Clara in a 16–6 victory. He punted eighteen times, with one punt measuring sixty-five yards. With his college eligibility ended after the 1941 season, he played in both the Shrine game and the College All-Star game.

Cleveland drafted Jacobs in the second round in 1942. He played for Cleveland in 1942, served in the U.S. Army in 1943–1944, and returned to play for Cleveland in 1945. In 1946, Jacobs played for the Washington Redskins and then finished his National Football League career with the Green Bay Packers, 1947–1949. In fifty-five NFL games, he attempted 552 passes and completed 244 for twenty-seven touchdowns. In 1948, he led the league in punting for Green Bay with sixty-nine punts for 2,782 yards, a forty-yard-per-punt average.

In 1950, he joined the Winnipeg Blue Bombers of the Canadian Football League to begin what Canadian sportswriters denote as a "decade of excitement." He earned All-Western and All-Star honors at quarterback, and led his team to the Grey Cup final game against the Toronto Argonauts. Toronto won 13–0. Winnipeg was back in the Grey Cup in 1953 against the Hamilton Tiger Cats, Jacobs having repeated as All-Western and All-Star quarterback. Even though Jacobs completed twenty-eight of forty-six passes for 326 yards, Winnipeg lost to Hamilton 12–6.

In 1973, Indian Jack Jacobs was inducted into the Canadian National Football League Hall of Fame. During his decade at Winnipeg, he rewrote western Canada records, completing 710 passes for 11,094 yards and 104 touchdowns. In 1977, he entered the American Indian Athletic Hall of Fame.

Royse Parr

FURTHER READING

Carroll, Bob, et al. *Total Football II.* New York: HarperCollins, 1999.

Clark, J. Brent. *100 Glorious Years of Oklahoma Football.* Kansas City, MO: Richardson, 1995.

Oxendine, Joseph B. *American Indian Sports Heritage.* Lincoln: University of Nebraska Press, 1995.

Clyde L. "Chief" JAMES

Born March 9, 1900
Died 1982
Basketball player

James (Modoc), an all-around athlete, earned a place in the American Indian Athletic Hall of Fame for his skills on the basketball court. As a forward for the Southwest Missouri State University Bears in the early 1920s, James set school and conference scoring records, and he continued to excel after college as an amateur and professional basketball player.

Son of Clyde S. James and Lydia Marvella Burns, an Irish immigrant, Clyde entered Seneca Indian School near Seneca, Missouri, at the age of seven. He played basketball in high school and then moved on to Southwest Missouri State in Springfield, Missouri, where he played in the backfield for the football team and showed promise as a baseball pitcher. Injuries prevented him from pursuing careers in either of those sports. Instead, he became one of the most memorable basketball players of his day.

At five feet, eleven inches and 170 pounds, James became a key element in the Bears basketball team's success, consistently averaging in double figures at left forward. During the 1924 season, he scored 175 points and helped lead the team to a Missouri Intercollegiate Association Championship title. James also gained election as captain of that year's All-League team. He gained great acclaim in local newspapers for his effective hook shot, "speedy" floor work, his record consistency as an "overhead"-style free-throw shooter (once hitting forty straight in six games), and his preference for hitting the "open ring" rather than trusting inconsistent backboards.

James continued with the game after college, leading the Missouri Valley Amateur Athletic Union in scoring field goals in 1925. As a forward for the Barnsdall "B-Squares" of Seneca, he gained selection as a tournament All-Star in 1926 and helped secure a team championship. He won another championship in 1927 with the Tulsa Eagles in Oklahoma. James stayed with that team, which later became the "Diamond Oilers," until 1947. Along with coach W.H. Bill Miller, James helped market the Mid-Continent Petroleum Corporation, which sponsored the team and employed many of its players, by winning games for the Oilers in competition throughout the Midwest. James also played for teams in Los Angeles and San Francisco, and spent time as an employee of the Bureau of Indian Affairs.

James took a job with the Bureau of Indian Affairs on the Navajo Nation, where he met Luella Mueller. They married in 1938 and moved to the Klamath Reservation in Oregon. He worked in a variety of careers, including managing a painting/screen-print gallery in Taos, New Mexico, during the 1950s and later managing the Wheeler County, Oregon Agriculture, Stabilization and Conservation Service. Luella died in 1962 and James in 1982. Both were buried in the Modoc Indian Cemetery, located near Miami, Oklahoma. A man of many accomplishments, athletic and otherwise, James gained the honor of induction to the American Indian Athletic Hall of Fame in 1977.

Wade Davies

FURTHER READING

James, Clyde file, American Indian Athletic Hall of Fame collection, Cultural Center and Museum, Haskell Indian Nations University, Lawrence, Kansas.

Oxendine, Joseph B. *American Indian Sports Heritage.* Lincoln: University of Nebraska Press, 1995.

Sydney I. JAMIESON

Born 1942
Lacrosse coach

Jamieson has coached the Bucknell University lacrosse team for more than thirty years. During this period he also has made numerous contributions to the sport at the national and international level.

Jamieson graduated from Cortland State University in 1964. Four years later, he was named head coach at Bucknell University. Over the next three decades, Jamieson proved himself to be a masterful coach. In 1996, Bucknell had an undefeated season, taking the Patriot League championship. As a result he was named the Patriot League Coach of the Year and USILA National Coach of the Year. Jamieson led the North team to victory in the 1998 North-South All-Star Game.

Jamieson has also played an important role in the development of modern lacrosse in indigenous communities. He has been active with the Iroquois Nationals team. He coached the team of Native Americans in the 1980s, taking it to the 1984 World Lacrosse Games, organized in association with the Los Angeles Summer Olympics. A year later the team toured England, playing ten matches against the English national team. In 1990, Jamieson led the team to the World Lacrosse Championships in Perth, Australia.

Jamieson has won numerous awards during his career. In 1985, he received the General George M. Gelston Award, recognizing the person who best symbolizes the game of lacrosse. Jamieson has twice won the Howdy Myers Memorial Award, honoring the "Man of the Year" in college lacrosse—first in 1986, and then again in 1996. He was also awarded the Burma-Bucknell Bowl, given for "outstanding contributions to intercultural and international understanding." Most recently, in 2003, he was named to the Pennsylvania Lacrosse Hall of Fame.

Jamieson and his wife Linda have raised three sons, Kevin, Steve, and Mark.

C. Richard King

George Howard JOHNSON

Born March 30, 1886, Winnebago Reservation
Died June 12, 1922, Des Moines, Iowa
Baseball player

Johnson was born on the Winnebago reservation, one of five children, two sons and three daughters, of Louisa Johnson. He learned to read and write at the Winnebago Agency school, but in his teen years he attended various federal boarding schools, including Lincoln Institute in Philadelphia, Carlisle Indian Industrial School in Carlisle, Pennsylvania, and Haskell Institute in Lawrence, Kansas. His Carlisle records describe Johnson as one-quarter-blood Winnebago and indicate that he arrived at the school in April 1900 and ran away that October.

In 1905, Johnson played outfield for a semipro team in Oakland, Nebraska, and he married another student from Carlisle, Margaret LaMere, a three-fifths-blood Winnebago. They were to have three children, Elaine, Catherine, and Joseph, while George played professional baseball during the summers and tended his homestead and barbered in Walthill, Nebraska, during the winters. He played his first professional ball with Guy Wilder Green's Nebraska Indians, a nationally recognized barnstorming team, in 1906 and 1907, and soon became the team's best pitcher. In 1907 Johnson pitched thirty-eight games for the Nebraska Indians and won thirty-two.

When Green bought a Lincoln Western League franchise, a Class A baseball team, in late 1907, he promptly signed Johnson to a contract.

For five years, Johnson toiled in the Western League, pitching for Lincoln, Sioux City, and St. Joseph, but did not achieve stardom until he learned to throw the spitball in 1910. In 1911, he threw a no-hitter against the Sioux City Packers, and in 1912, he recorded a 24–12 record and attracted the attention of major league clubs. He was signed by the White Sox but was traded to the Cincinnati Reds in early April 1913. He made a spectacular major league debut on April 16, 1913, at Redland Field, when he shut out the St. Louis Cardinals on three hits. Johnson went on to lead the Reds in most pitching categories that season. He signed a contract with Gary Herrmann's Reds for the 1914 season, but in spring training and early in the season, he was fined heavily for violation of training rules by new manager Buck Herzog.

Courted by the Kansas City Packers of the upstart Federal League, Johnson jumped his National League contract and signed with the Kansas City team in April 1914. For more than three months of the season, Johnson was prevented from pitching outside of batting practice by lawsuits filed by the Reds. He joined the rotation in August and threw two shutouts in September. In 1915, Johnson rebounded and had his best season in major league baseball, pitching in forty-six games with seventeen wins, four shutouts, and a 2.75 ERA.

When the Federal League folded at the end of the 1915 season, Johnson returned to the minor leagues, pitching for three years in the Class AA Pacific Coast League, until he injured his shoulder severely in 1918. He made a comeback attempt in early 1919 with the Dallas Marines in the Class B Texas League, but the shoulder injury was irreparable. George Johnson was working as a traveling salesman in 1922 when his life ended tragically. In the early hours of June 12, he was shot and killed during a dice game in Des Moines, Iowa.

Johnson gained national sporting headlines as a major league pitcher from 1913 to 1915, pitching nine big-league shutouts and two minor-league no-hitters. Nicknamed "Chief" and heckled by fans for his American Indian identity, Johnson pitched with determination and success, enjoying a fourteen-year professional career.

Jeffrey Powers-Beck

FURTHER READING

Powers-Beck, Jeffrey. "'Winnebago is a Great Nation': George Howard Johnson's Life in Baseball." In *Telling Achievements: Native American Athletes in Sport and Society*, edited by C. Richard King. Lincoln: University of Nebraska Press, forthcoming.

James JOHNSON

Born June 6, 1879
Died January 1942
Football player

Johnson (Stockbridge-Munsee) replaced Frank Hudson as Carlisle's quarterback in 1899 and played until 1903. During Johnson's five years at Carlisle, the team posted a 39–18–3 record. Johnson earned All-American honors in 1901 and 1903.

In 1905, Johnson married Florence Welch (Oneida). He attended Dickinson College and later Northwestern's Dental School. After earning his dentistry degree, Johnson opened a practice in Puerto Rico and school officials hailed Johnson as the exemplar of the school's mission to assimilate Native Americans. He later married a woman from Puerto Rico and had one daughter.

Coaches and journalists remembered Johnson as a shrewd quarterback and an excellent kicker. "Pop" Warner named Johnson to the All-Time Carlisle Indian team at quarterback, citing his "masterful leadership, strategic ability, and physical prowess." Johnson was an agile runner and possessed an accurate arm. Johnson passed away in 1941. He was inducted into the inaugural class of the American Indian Athletic Hall of Fame in 1972.

William J. Bauer, Jr.

FURTHER READING

Adams, David Wallace. "More than a Game: The Carlisle Indians Take to the Gridiron, 1893–1917." *Western Historical Quarterly* 32 (Spring 2001): 25–54.

Newcombe, Jack. *The Best of the Athletic Boys: The White Man's Impact on Jim Thorpe*. Garden City, NY: Doubleday, 1977.

Oxendine, Joseph. *American Indian Sports Heritage*. Champaign, IL: Human Kinetics, 1988.

Steckbeck, John. *Fabulous Redmen: The Carlisle Indians and Their Famous Football Teams*. Harrisburg, PA: J. Horace McFarland, 1951.

John Henry JOHNSON

Born August 21, 1956, Houston, Texas
Baseball player

John Henry Johnson was a tall left-handed pitcher who had a marginal major league baseball career in the 1980s for the Oakland As, Texas Rangers, Boston Red Sox, and Milwaukee Brewers. In between stints in the minors, Johnson appeared in 214 major league games, pitching 602.7 innings. In eight years, he had a respectable 3.90 era and a record of 26–33.

Born on August 21, 1956, in Houston, Texas, Johnson came to the major leagues after the Oakland As' owner, Charlie Finley, tore apart his 1970s championship teams for financial reasons. Johnson came to the As as part of a seven-for-one trade for pitcher Vida Blue in 1978. His major league career started with a bang on April 10, 1978, after which he led the team in wins, with eleven. Johnson was named to several all-Rookie teams, and great things were predicted for him. In 1979, however, he struggled with the "sophomore jinx," going 2–8, with a 4.36 ERA in fourteen games.

Having little patience with his changing fortunes, the As traded the twenty-two-year-old to the Texas Rangers in June. While Johnson was happy about the trade, the change in scenery did not change his fortunes. Johnson went 2–6 in seventeen games, and his ERA increased to almost five. In 1980, believing that his fastball would do better out of the bullpen, the Ranger moved him into middle relief. In his new role he did a decent job over the next two seasons, posting a 2.33 and 2.66 ERA, respectively, in fifty-seven games. Johnson signed with the Boston Red Sox as a free agent in 1982 and spent the entire season in the minor leagues. He resurfaced in the Boston bullpen the next season and was a mediocre workhorse for the Red Sox the next two years, posting ERAs of 3.71 and 3.53 in sixty-four games.

Again, Johnson did not live up to expectations that management had for him. Returning to the minor leagues in 1985, he pitched a no-hitter on May 2, 1985, for Hawaii of the Pacific Coast League against Calgary. He resurfaced in the major leagues with the Milwaukee Brewers in a brief stay in the 1986. He appeared in ten games for them in 1987 but was pitiful in his 26.3 innings, posting a 9.57 ERA before the team decided to cut him. Unable to hook on with another team, Johnson retired.

T. Jason Soderstrum

FURTHER READING

Bosetti, Rick. "Rookie All-Star Team Paced." *Sporting News*, December 9, 1978, 43.

Galloway, Randy. "'Leaving As Just Like a Pardon,' Says Johnson." *Sporting News*, July 7, 1979, 5–6.
Vass, George. "Baseball Digest's 1978 Rookie All-Star Team." *Baseball Digest* 37 (November 1978): 18–27.
Weir, Tom. "As' Johnson Proud of Indian Blood." *Sporting News*, May 13, 1978, 20.

Robert Lee "Indian Bob" JOHNSON

Born November 26, 1905, Pryor Creek, Oklahoma
Died July 6, 1982, Tacoma, Washington
Baseball player

Robert Lee Johnson, of Cherokee descent, was one of eight children. His formal education likely ended in grade school, when his family moved to the state of Washington. During his adolescence he began an itinerant lifestyle, one that took him to Glendale, California, to work for the fire department. Along the way, he married at age eighteen Caroline Stout in 1924 and quickly had two daughters, Roberta Louise and Beverly Jean.

Johnson followed his older brother Roy Cleveland into professional baseball, playing first in the Western International League, followed by a full season with Portland of the Pacific Coast League (PCL) in 1930. Three years later, he joined the Philadelphia As, initiating a thirteen-year career as an outfielder in the major leagues that took him to the Boston Red Sox and Washington Senators as well. During his playing days, Johnson displayed extraordinary year-to-year consistency, albeit with virtually no league-leading performances. Johnson was a seven-time All-Star selection, noteworthy for his versatility and power hitting (.393 on-base percentage and .506 slugging percentage).

Johnson, nicknamed "Indian Bob," "Cherokee Bob," and "Kickapoo," initially promoted his identification as an Indian ballplayer. Later in his career, he disavowed his Indian heritage, stating (erroneously) that his mother was only one-sixteenth Indian, thereby making him merely one-thirty-second Indian. Undoubtedly, he had grown tired of the constant "wahooing" and Indian pidgin language used in reference to him by fans and sportswriters. Even fans of the As in the Shibe Park left-field bleachers would roar the Indian war whoop whenever Johnson came up to bat.

Following the close of his major league career in 1945, he returned to the minor leagues, as did many players of the time. Johnson continued to play intermittently until 1951, and he also managed in the minor leagues. He was understood to be interested in a major league managerial career, but this never came to fruition. Instead, he returned to his adopted hometown of Tacoma, working for a beer distributorship, remarrying and having a namesake son, and ultimately dying at the age of seventy-six. Sadly, Johnson never gained recognition from the Hall of Fame selectors, receiving only two votes, one in 1948 and the other in 1956.

Doron Goldman

FURTHER READING

Thorn, John, et al., eds. *Total Baseball: The Official Encyclopedia of Major League Baseball.* 5th ed. New York: Viking, 1997.

Roy Cleveland JOHNSON

Born February 23, 1903, Oklahoma Territory
Died September 10, 1973, Tacoma, Washington
Baseball player

Johnson (Cherokee) was one of eight children of Anna Blanche Downing (or Dirtthrower) and Robert Lee Johnson. He moved with his family to Tacoma, Washington, while in grammar school. At the age of nineteen, he began to play semiprofessional ball with Everett, Washington. At first, Johnson played both as a pitcher and an outfielder, but he eventually signed with San Francisco of the Pacific Coast League as an outfielder. Roy did not play full-time for San Francisco until 1928, when he had a breakout season, batting .360, with twenty-two home runs, twenty-nine stolen bases, and 142 runs scored.

In Johnson's rookie season with Detroit, 1929, he batted .314 with 201 hits, still fifteenth all-time by a rookie. He also set a modern American League record by making thirty-one errors in the outfield. In mid-1932, the Tigers traded him to the Boston Red Sox, who in turn traded him to the Washington Senators at the end of the 1935 season; they in turn dealt him to the New York Yankees a month later. Johnson finished his career in 1937 and 1938 with the Boston Braves, ending up with a career .296 batting average, the same as his younger brother, Robert Lee "Indian Bob" Johnson. Roy was known as a fine outfielder with a strong throwing arm and considerable speed, as evidenced by his career total of 135 stolen bases, a significant total for those times.

Johnson married Helen Lucille Fraser on October 25, 1929, and had one daughter, Marilyn. The couple later divorced. Johnson died back in Tacoma on September 10, 1973.

Doron Goldman

FURTHER READING

Porter, David L., ed. *Biographical Dictionary of American Sports: Baseball*. Westport, CT: Greenwood, 1987.

Reichler, Joe. *The Baseball Trade Register*. New York: Macmillan, 1984.

Roy Johnson obituary, *Tacoma News Tribune*, September 11, 1973.

Thorn, John, Pete Palmer, Michael Gershman, Matthew Silverman, Sean Lahman, and Greg Spira, eds. *Total Baseball*. 5d ed. New York: Viking, 1997.

Stanley Carl "Bulldog" JONATHAN

Born May 9, 1955, Ohsweken, Ontario
Hockey player

Stan Jonathan played left wing for the Boston Bruins for over a decade. Although he is often remembered by fans for his toughness and pugilism, Jonathan was a valuable all-around contributor to Bruins clubs that made two Stanley Cup finals (1977–1979).

A full-blooded Tuscarora, Jonathan grew up on the Six Nations Reserve near Brantford, Ontario, where he returned every summer after he turned professional. His career began in 1972 with the Peterborough Petes of the Ontario Hockey League. In 1973–1974, Jonathan also played in the unofficial first World Junior Championship in Leningrad. His Canadian national squad finished third.

Jonathan improved steadily during three years in the juniors, scoring thirty-six goals with thirty-nine assists in 1974–1975. The Bruins made him their fifth-round pick (eighty-sixth overall) in the 1975 amateur draft. Stan spent most of the 1975–1976 season with the Dayton Gems, who won the International Hockey League championship, led by Jonathan's thirteen goals and eight assists in fifteen playoff games.

He made his NHL debut on April 4, 1976. Jonathan was with Rochester of the American Hockey League for three games in 1976–1977, but he made the big club to stay that season. He led the NHL in shooting percentage, as seventeen of his seventy-one shots on goal lit the lamp. The Bruins finished first in their division and marched through the first two rounds of the playoffs before the Montreal Canadiens swept them in the Stanley Cup finals.

Jonathan's greatest NHL success came in 1977–1978, notching a career-high twenty-seven goals and twenty-five assists. Again Boston won the first two playoff rounds but lost the Stanley Cup to the Canadiens. The Bruins ran into a roadblock once more in Montreal the next year. In game six of the conference finals, Jonathan (out most of the year with a separated shoulder) enjoyed probably his peak moment. He scored a hat trick at home, sending the series to a truly tense game seven at the old Montreal Forum. Unfortunately, Boston drew a costly late penalty, and Montreal converted the power play, winning in overtime.

Jonathan scored twenty-one goals in 1979–1980, but then injuries, a consequence of his physical style, took a mounting toll. The Bruins sold his rights to Pittsburgh on November 8, 1982, and he played the bulk of his last season in the AHL. A couple of years later, Stan made a recreational comeback in the Ontario Senior League. Since retiring, he has been active in Aboriginal causes—notably, youth hockey programs such as those sponsored by Algonquin Gino Odjick, who named Jonathan his boyhood hero.

Rory Costello

FURTHER READING

Fischler, Stan. *Boston Bruins: Greatest Moments and Players.* Champaign, IL: Sports Publishing, 2000.

McFarlane, Brian. *The Bruins: Brian McFarlane's Original Six.* Toronto: Stoddart, 1999.

Frank JUDE

Born 1884, Libby, Minnesota
Died May 4, 1961, Brownsville, Texas
Baseball player

Frank Jude was born in 1884 in Libby, Minnesota. A Chippewa, Jude attended the Carlisle Indian School. He made his major league baseball debut at age twenty-two on July 9, 1906, with the National League's Cincinnati Reds. He played for one season on that one team and ended his big-league playing career in 1906. Jude was a right-handed player who weighed 150 pounds and stood five feet, seven inches tall. In his only season, he played in eighty games as a right fielder and had 308 at bats. He had sixty-four hits, six doubles, four triples, and one home run, for eighty-one total bases. He scored thirty-one runs and batted in sixteen runs. Jude died on May 4, 1961, in Brownsville, Texas.

Larry S. Bonura

FURTHER READING

Baseball Almanac (www.baseball-almanac.com/players/player.php?p=judefr01).

Baseball-Reference.com (www.baseball-reference.com/j/judefr01.shtml).

CNN/Sports Illustrated (http://sportsillustrated.cnn.com/baseball/mlb/all_time_stats/players/j/48209).

Thorn, John, and Pete Palmer, eds. *Total Baseball: The Ultimate Encyclopedia of Baseball.* 3rd ed. New York: HarperPerennial, 1993.

Isaac Leonard "Ike" "Chief" KAHDOT

Born October 22, 1901
Died 1999
Baseball player

Ike Kahdot, a Potawatomi, was for many years the oldest living member of major league baseball's Cleveland Indians team. He was the first Native American to play for Cleveland since outfielder Louis Sockalexis (1897–1899) and the last until a fellow Oklahoman, pitcher Allie Reynolds (1942–1946) signed with the team.

When he was growing up, Kahdot played on a team of boys in his mostly Indian village. Later he starred at shortstop on the high school team at Haskell Institute in Lawrence, Kansas. After graduating from Haskell, he was hired to play baseball for Empire Oil and Gas Company's semipro team in Bartlesville, Oklahoma, where he played in 1919 and 1920. In 1921, Kahdot played minor league baseball in the Southwestern League for Pittsburg, Kansas, where he hit .322. In 1922 while playing for Coffeyville, Kansas, he led the same league with 111 runs.

At the end of the season, Coffeyville sold his minor league contract to the Cleveland Indians. On September 5, 1922, he made his major league debut in the sixth inning as a pinch runner against the St. Louis Browns. The next day he made his first plate appearance and grounded out as the Browns completed a four-game sweep. Kahdot made limited appearances in two other games and went hitless. As the season wound down, he once shared a chew of tobacco with Babe Ruth of the New York Yankees and met several future National Baseball Hall of Famers, including his manager Tris Speaker and Ty Cobb of the Detroit Tigers.

In the Indians clubhouse after the last game of the season, manager Speaker passed around baseballs for everyone to sign. Each player got to take one home. This was his only souvenir of his season, a lifelong treasure for Kahdot. He left for Kansas and would never return to Cleveland.

For the 1923 season, the Cleveland Indians asked Kahdot to move to Grand Rapids, Michigan, to join a team to which they commonly sent promising players. By then, however, Kahdot had married and set his family roots in Coffeyville, Kansas. A possible major league career did not seem worth the move. He never regretted his decision not to go to Grand Rapids. Kahdot played for various minor league and semipro teams until he retired from baseball in 1941. By then he had moved to Oklahoma City and was working in the Oklahoma oil fields.

In the off-season Kahdot developed a close friendship with Coffeyville's most famous resident, Walter Johnson, one of the greatest National Baseball Hall of Fame pitchers ever to grip the seams. They enjoyed hunting together with their coon dogs. Kahdot often played shortstop on exhibition teams that Johnson formed for barnstorming tours in Kansas and Oklahoma.

Royse Parr

FURTHER READING

Burke, Bob, Kenny A. Franks, and Royse Parr. *Glory Days of Summer: The History of Baseball in Oklahoma.* Oklahoma City: Oklahoma Heritage, 1999.
Spencer, Burl. "'Chief' Was an Indian in Ruthian Age of Dreams." *Tulsa World*, September 22, 1993.

KICK STICK

Stick games of varying type and difficulty were common among many Native Americans. Some games involved chance or dexterity, but others resembled modern sports. Those often used sticks to hit balls, as in baseball, or as the main implement instead of a ball, as in soccer. Stick games familiar to many Americans include double ball, lacrosse, and stick ball or toli. Kick stick is less well known but no less deserving of attention.

A variation of a ball race, Native Americans played the strenuous and highly competitive game of kick stick with various types of decorated billets of wood. Sticks, which usually had intricate designs for identification by owners and judges during and after races, averaged three-inch lengths and one-inch widths. The Zuñi, who used longer sticks, thought that the kick stick kept players running as long as the stick traveled ahead. The Pima also believed they could run faster if they ran while kicking a ball or stick. Races usually consisted of two men or teams of three to six players. Zuñi runners held on to favorite kick sticks and considered them to be endowed with magic. Magic also had a place in races among the Maricopa and Pima. This element of magic probably resulted from the difficulty of races, since sticks could be touched only with bare feet, lifting them into the air from underneath with the instep. Races covered several miles of uneven terrain with potential stick traps, such as streams or sand.

Kick stick was the most popular outdoor game among Hopi, who often played in teams of four or five runners on cross-country courses of twenty to thirty miles. The Zuñi similarly played in male teams of varying ages on twenty-five-mile courses. Good kickers sent a stick about twenty to thirty yards. More than a popular game, Zuñi kick stick races *(ti-kwa-we)* were often important ceremonial events to bring rain; the sticks were buried in a cornfield after races. Kick stick was associated with the War Gods. Kick sticks were also emblematic of swift journeys and the Twin Gods' miniature bows that pursued and were pursued by men in contention. Although races often left players sore and injured, men practiced all year. Zuñi practiced eight months of the year before the championship, which was of equally great interest to women. Hopi races were also related to rain. Like most Zuñi races that took place in the spring between the planting of wheat and corn, Hopi races occurred in spring and early summer. The Hopi believed that rolling the ball or kicking the stick initiated streams of water down gullies and canyons. A more modern version was played on the Six Nations Reservation in the Woodlands region, with crescent-shaped sticks ten inches long.

Diana Meneses

See also: Ball Race; Running.

FURTHER READING

Baldwin, Gordon C. *Games of the American Indian*. New York: Grossest and Dunlap, 1969.
Blanchard, Kendall. *The Anthropology of Sport: An Introduction*. Westport, CT: Bergin and Garvey, 1995.
Fletcher, Alice C. *Indian Games and Dances with Native Songs*. Boston: C.C. Birchard, 1916.
MacFarlan, Allan, and Paulette Macfarlan. *Handbook of American Indian Games*. New York: Dover, 1985.
Owens, John G. "Some Games of the Zuñi." *Popular Science Monthly* 39 (May 1891): 39–50.
Oxendine, Joseph B. *American Indian Sports Heritage*. Lincoln: University of Nebraska Press, 1995.

Darris KILGOUR

Born September 21, 1970, Niagara Falls, New York
Lacrosse player

Kilgour was born to Richard and Christine Kilgour. In 1988, he graduated from

Niagara-Wheatfield High School, where he played lacrosse and basketball.

Kilgour played amateur lacrosse with five different clubs in the Ontario Lacrosse Association (OLA). He began with Niagara-on-the-Lake (1987–1988). Next, he joined the St. Catharine's Athletics (1989–1991), winning the Advertiser Trophy by leading the league in scoring in 1990 and 1991. He also led his team to Minto Cup championships both seasons. In 1991, Kilgour won the Dennis McIntosh Memorial Trophy as the most valuable player in the championship series. Kilgour then played for the Brampton Excelsiors of the OLA's Major Series (1992–1993), winning the Mann Cup, the senior championship of Canada, both seasons. Kilgour was fifth in the OLA in scoring in 1993.

From 1994 to 1996, Kilgour played for the Six Nations Chiefs and won the Mann Cup each year. In 1994, he won the Mike Kelly Memorial Trophy as the most valuable player in the championship series. In 1997, he played for the Niagara Falls (ON) Gamblers and remained with the club for the 1998 season, when the team was relocated to Buffalo, NY. He concluded his amateur playing career in 2000 with the Brooklin Redmen.

Professionally, Kilgour played for the Buffalo Bandits from 1992 to 1998, winning Major Indoor Lacrosse League championships in 1992, 1993 and 1996. In the middle of the 1998 season, Kilgour was traded to the Rochester Knighthawks. He finished his professional playing career with the Albany Attack in 1999. The Buffalo Bandits honored Kilgour's service with the club by retiring his jersey, number forty-three.

He became only the second Native head coach in the National Lacrosse League (formerly known as the MILL) when he led the Washington Power for the 2001 season. He remained with the team through 2002 but stepped down when the franchise moved to Denver. He continued his head coaching career with the Buffalo Bandits for the 2003 season. Kilgour was named coach of the year by the National Lacrosse League in 2003.

Donald M. Fisher

FURTHER READING

Fisher, Donald M. *Lacrosse: A History of the Game*. Baltimore: Johns Hopkins University Press, 2002.

Rich KILGOUR

Born January 14, 1969, Niagara Falls, New York
Lacrosse player

Kilgour was born to Richard and Christine Kilgour. In 1987, he graduated from Niagara-Wheatfield High School, where he played lacrosse, football, and basketball. He also played lacrosse at Nazareth College, graduating in 1991.

Kilgour played his amateur lacrosse career for five different clubs in the Ontario Lacrosse Association (OLA). He played four seasons with the St. Catharines Athletics from 1987 to 1990. In 1990, the Athletics won the national junior championship of Canada by winning the Minto Cup in 1990. In 1991, he moved up to the Major Series of the OLA and played for the Brampton Excelsiors. Kilgour also played with the Excelsiors in 1993 and helped the team win the Mann Cup, the senior championship of Canada.

From 1994 to 1996, Kilgour played for the Six Nations Chiefs and won the Mann Cup all three years. In 1997, he played for the Niagara Falls (ON) Gamblers, who lost to the Victoria Shamrocks in the Mann Cup finals. Kilgour was fourth in the OLA in scoring. The following season, when the club relocated to Buffalo, NY, Kilgour also served as the club's general manager. For the 2000 and 2001 seasons, he played for the Akwesasne Thunder.

Kilgour began his professional lacrosse career in 1991 with the Buffalo Renegades of the short-lived National Lacrosse League. In 1992, Kilgour became an original member of the Buffalo Bandits of the professional Major Indoor Lacrosse League. The Bandits won championships in 1992, 1993 and 1996, all against the Philadelphia Wings. For the 2003 season, Kilgour played his twelfth season with the Bandits.

Donald M. Fisher

FURTHER READING

Fisher, Donald M. *Lacrosse: A History of the Game*. Baltimore: Johns Hopkins University Press, 2002.

Wayne KING

Born September 4, 1951
Hockey player

Prior to moving to Niagara Falls to play junior hockey at the age of sixteen, Wayne King (Ojibwa) was just another young hockey player from small-town Canada. Both his parents had grown up on Ontario reserves and moved into what they considered a large center in search of work. Nevertheless, the move from the country to the city proved an exciting experience for young Wayne, a hockey player who would one day call California and the NHL his home.

King grew up in the tiny town of Fort McNicoll, where his family was the only Native family in town. His father worked as an engineer and sailor, while his mother stayed at home with the children. King showed a talent for hockey, and at age fifteen he was invited to try out for the Niagara Falls Flyers of the Ontario Hockey League (OHL). Although he would not make the team until the next year, he played with the Flyers for three years and eventually was drafted by an NHL expansion team, the California Golden Seals. He was then assigned to the Columbus Golden Seals for one season and the Salt Lake Golden Seals for two more before making the parent club in 1974.

King played defensive forward. He missed a significant portion of his rookie season with a knee injury, later compounded by a burst appendix that forced him out of even more action. All told, King appeared in only twenty-five games his inaugural season. He appeared in forty-six games the following season, which would prove his final year in the NHL following the Golden Seals announcement in January 1976 that they would cease operations. King was left without a team at season's

Wayne King had a short professional hockey career with the Salt Lake City Seals in the 1970s. *(Hockey Hall of Fame and Museum)*

end; when the remaining pro clubs expressed little interest, he chose to retire at the end of 1977 after playing a final year with the Salt Lake Golden Seals. King finished his NHL career at the age of twenty-six with five goals and twenty-three points in seventy-three games.

Upon his return to Midland in 1977, King worked for the next two years in an auto body shop while playing one year in the OHL Senior Division. He also began his second career when he secured work as a mental health worker/security guard at the Penetanguishene Mental Health Centre, attending two years of nursing courses while obtaining his Registered Practical Nursing designation.

It was during this period that King and his wife, who remained with her husband during his myriad travels through the professional ranks, decided to start a family; a son and daughter soon followed. Today King still works for the Ontario government and spends as much of his free time as possible on the golf course.

Yale D. Belanger

FURTHER READING

The Hockey Data Base (www.hockeydb.com/ihdb/stats/pdisplay.php3?pid 871).

"Wayne King" (www.legendsofhockey.net:8080/LegendsOfHockey/jsp/SearchPlayer.jsp?player=13197), accessed October 30, 2003.

John KORDIC

Born March 22, 1965, Edmonton, Alberta
Died August 8, 1992, Quebec City, Quebec
Hockey player

Kordic grew up in Edmonton and started playing hockey at an early age. He was known as a shy, timid child. After he started playing professional hockey with the Portland Winter Hawks of the Western Hockey League in the early 1980s he first displayed the aggressive, violent style of play that later became his trademark. A talented player, Kordic was one of the highest-scoring defensive players in the league, shooting twenty-three goals with fifty-eight assists for the Winter Hawks and the Seattle Breakers in the 1984–1985 season. But he was increasingly pressured by his coaches to adopt the role of "enforcer," using fighting and other violent means to intimidate or disable opponents. He also found that he enjoyed the attention this role brought.

Impressed with his fighting ability, the Montreal Canadians called him up near the end of their 1985–1986 season, in which they won the Stanley Cup. He played three more seasons with the Canadians, never managing more than eight points in any season. His role was clear: he became one of the league's most notorious "goons." Kordic was originally a skinny man, but he started taking steroids and bulked up to 220 pounds by the start of the 1986–1987 season. Kordic was deeply conflicted by his desire to succeed in the NHL no matter what it took, his own dislike of fighting, and his father's strong disapproval of his adopted style of play. The stress lead him to begin using cocaine and to drink heavily.

When Kordic was traded away in 1988 to the Toronto Maple Leafs, his troubles increased. He was now regarded as a player of limited skills brought in for only one purpose, fighting. In his second season with the Maple Leafs, he showed he still had some ability, scoring a career-high nine goals, but he did not enjoy the same popularity with fans he had in Montreal, and he had fights with his own teammates. His father's death in October 1989, was devastating to Kordic; he was never the same afterward. His cocaine abuse escalated, and he was cut by the team in 1990. He signed with the Washington Capitols in February

1991 but played in only seven games. He was twice suspended for alcohol-related offenses and ended up in a substance-abuse treatment center. He was released by the Capitols in June 1991.

He signed with the bottom-dwelling Quebec Nordiques for the 1991–1992 season but could not control his addictions and was released in January 1992. After a short stint with the Cape Breton Oilers, he returned to Quebec City. His many attempts to halt his drug addictions failed. He died of heart failure during a drug-induced rage so violent that it brought the police to his hotel room.

Edward W. Hathaway

FURTHER READING

Scher, Jon. "Death of a Goon." *Sports Illustrated,* August 24, 1992.

Zwolinski, Mark. "The John Kordic Story: The Fight of His Life." Toronto: Macmillan Canada, 1995.

LACROSSE

Lacrosse is not only the oldest team sport in North America today, but it is also one of the few forms of Native culture surviving into the early twenty-first century with appeal among non-Native peoples. The name "lacrosse" is a generic label referring to several different versions of what used to be one game. However, all feature two teams of players using webbed sticks to carry, pass, and shoot a ball into goal nets at opposite ends of an oblong rectangular playing surface. There are two distinct versions of field lacrosse in the United States: the men's game has ten players to a side, outfitted in helmets and shoulder pads, while the women's game has twelve players wearing no significant protective gear. Limited forms of stick-on-body and body-to-body contact is permitted in the male game, but the women's version minimizes it. In Canada, box lacrosse features only six players on a smaller playing surface, either on an enclosed, outdoor field or inside, in a converted hockey arena. Unlike field lacrosse, this version of the sport features greater physical contact, and walls serve as boundary lines. The sport of indoor lacrosse, a modified version of box lacrosse, is currently played by a professional league. A more recent creation, intercrosse, is a game for children deemphasizing competition and roughness. Most lacrosse sticks today are made from synthetic materials, but Native wood craftsmen continue to produce sticks as well.

Origins of the Game

Contemporary versions of lacrosse evolved from a sport played by middle-class Canadian nationalists living in Montreal during the mid-nineteenth century. These white sportsmen had learned the game from Mohawk Indians from the nearby Caughnawaga (Kahnawake) Reserve. The game played by the distant forefathers of those Indians was in turn part of a diverse tradition of stick-and-ball games played throughout eastern North America for generations. The European missionaries, explorers, and other travelers who observed Native contests during the seventeenth, eighteenth, and nineteenth centuries identified at least three distinct regional clusters of stick-and-ball games. Nations living around the eastern Great Lakes and Saint Lawrence River valley played a game utilizing a wood stick with an unenclosed pocket; other tribes, living to the west, around the western Great Lakes and upper Mississippi River drainage system, used a shorter stick with an enclosed pocket; and still other peoples, who lived around the lower Mississippi River drainage system and southern Atlantic coastal plain, played a game with two sticks with enclosed pockets.

Current explanations of the origins of these games mirror larger debates between science and religion on the origin of life. According to historians and anthropologists, it is unknown whether these tribal games arose independently, originated in one region and spread elsewhere, or perhaps had very ancient origins in the sophisticated ball-court games of Mesoamerica. Meanwhile, Native religious traditionalists contend the game was a divine gift from the Creator or from nature itself. Team sizes, playing surfaces, equipment, and the temporal and cultural contexts of Native matches varied. Two teams, ranging anywhere from a small number of participants to hundreds, fought for possession of a ball in order to throw it at a designated goal marker. Sometimes reli-

An artistic rendering of a traditional lacrosse match. *(Rochester Museum)*

gious leaders called for the playing of a contest to heal ailing members of the community or to honor individuals who had died. Not only did ball play keep men in shape for military and hunting activities, but it resolved disputes among neighboring allied peoples. Spectators wagered over the outcome of contests, while participants took advantage of a game's rough play to settle old scores with personal rivals.

When French Jesuit priests first encountered these Native games, during the seventeenth century, they referred to them collectively as "la crosse," a general term referring to any playing stick with a curved end. Until the mid-eighteenth century, few white men knew much at all about the ballgame traditions of North America. During the aftermath of the French surrender of Canada to Great Britain in 1760, Indians throughout the Great Lakes and Ohio River Valley arose in opposition to the new British occupation of old French forts. Ojibwa and Fox Indians staged a contest in front of the garrison of Fort Michilimackinac in the northern part of modern Michigan in 1763. Merely a ruse, the game provided cover for the eventual assault on the unsuspecting British troops. For many English-speaking colonists to the east, the attack reinforced negative attitudes of Indians as ignoble savages.

Canada's "National Game"

The modern sport of lacrosse was born around the time of the confederation of Canada in 1867. A young dentist named William George Beers learned lacrosse from Mohawk Indians and urged other white sportsmen in Montreal to embrace the virtues of his adopted Indian game. In Beers's mind, lacrosse could teach Canadian men valuable cultural values and mental skills. Throughout the 1860s, Beers codified the game by writing and revising rules. By reducing playing fields to two hundred yards from goal to goal, Beers made the game more compatible with the spatial limitations of the modern city. Indian games had once been played on large open fields in the countryside, and they emphasized running ability over passing technique. Bowing to complaints that Indians grabbed the ball and then outran white players over these large fields, Beers introduced the new short, standardized fields to negate the perceived Indian advantage. Beers also outlined the principle of positional play by limiting an on-field contingent to twelve players, each with his own specific title and role. Beers's game had much in common with other team games experiencing modernization in England, namely soccer and rugby football. However, since lacrosse proponents learned the game from the continent's indigenous people, Beers believed it was a more authentic game for the new Canadian nation.

Now propagandized as the country's "national game," lacrosse diffused from Montreal throughout Quebec and Ontario, and from Anglos to the Irish and French. The rise in the popularity of the sport led to play between clubs from different cit-

Players from Syracuse University and the University of Virginia square off in an intercollegiate match. *(AP/Wide World Photos)*

ies, the establishment of championship challenge contests, the collection of gate fees from spectators, organized gambling, the covert hiring of expert "professional" players (especially Indians), accusations of bad officiating and deliberate on-field violence, longer rule books, and formal leagues. Essentially, what had briefly been a "gentleman's game" had been turned into a commercial sport propagated by a community's business elite. By the 1880s, intense competition among Canada's many clubs led teams to bolt from one organized league to another. Controversies over professionalism also caused some white organizations to ban Native participation.

From the 1860s through the 1920s, the game spread not only from middle-class, English-speaking Canadian men in Montreal to other cities and social classes in Canada, but also to private athletic circles in the United States and Great Britain. Especially important in the diffusion of lacrosse to the United States and Britain were Mohawk Indian teams from the Caughnawaga and Saint Regis (Akwesasne) Reserves. Often playing against other reservation teams or against white clubs in Canada or the United States, these Indian tour teams helped to promote lacrosse in the greater New York City area, Boston, and Newport, Rhode Island. Curious East Coast spectators were able to witness flesh-and-blood Indians playing what everyone now knew to be a white man's game learned from Indians. Indian and "gentleman" teams from Canada also made trips to the British Isles during the 1870s and 1880s. These clubs wore colorful costumes that highlighted racial differences. Some Canadian proponents of lacrosse hoped the game would help encourage Englishmen to emigrate to Canada. Meanwhile, not only did aristocratic Englishmen adopt lacrosse and establish their own clubs, but women's physical educators in Britain developed a tamer version of lacrosse for schoolgirls by the 1890s.

Cultural Folkways of Twentieth-Century Lacrosse

During the years prior to the First World War, lacrosse in Canada was staged as a commercial sport by semiprofessional clubs. However, problems marred the game: escalating player salaries, insufficient gate revenue, on-field violence, and controversial officiating. By the 1920s, pro lacrosse had fallen into disfavor, and many Canadians chose instead to support baseball and softball. The surviving amateur lacrosse clubs operated as best they could, but the sport had clearly lost its old prominent position on the summer sports calendar. Meanwhile, commercial ice hockey promoters tried to capture the public's interest by creating a professional indoor version of the game, called "box lacrosse," in 1931. The league failed to finish its second season, but amateur field lacrosse clubs and governing bodies across the country voted to switch to a box lacrosse format the following year.

Canadian immigrants and tour teams played a large role in the early development of organized lacrosse in the United States during the 1870s and 1880s. The formation of dozens of private athletic clubs was followed by the adoption of the game by northeastern colleges. During the early decades of the twentieth century, Baltimore and Brooklyn became homes to some of the most successful lacrosse teams on university campuses, at private clubs, and even in preparatory and secondary schools. The lacrosse team at Johns Hopkins University earned a reputation as consistently one of the best in the country. Ironically, the sport devised by Beers now resided in the United States. In 1971, the NCAA began sponsoring a national championship tournament. More recently, Syracuse and Princeton universities have established dominant programs.

Beginning in the 1930s, efforts were made to popularize women's lacrosse in the United States. English-trained American women who studied physical education overseas brought back the new genteel brand of the game. Clubs in and around Baltimore, Philadelphia, and New York served as the basis of a new American women's lacrosse community. The women's game spread slowly until the 1980s, when the numbers of intercollegiate and interscholastic programs multiplied rapidly. The University of Maryland has consistently fielded especially strong teams.

Native peoples participated in both American field lacrosse and Canadian box lacrosse during the twentieth century. During the first half of the century, reservation teams representing the Onondaga, Saint Regis (Akwesasne), Caughnawaga, Cattaraugus, Akron (Tonawanda), and Six Nations communities played numerous exhibition contests against such collegian teams as Syracuse University and Hobart College, as well as private athletic clubs. A few Indians were even recruited to play for university teams. For instance, Victor Ross earned All-American honors playing for Syracuse in 1923. After World War II, contests between reservations and universities became less common. The Carlisle Industrial Indian School also fielded a lacrosse team during the second decade of the century. Even though Ivy League schools and Johns Hopkins usually dominated the sport, the Carlisle Indians were often regarded as having one of the best teams in the country. Many of Carlisle's athletes came from the reservations in New York.

During the Great Depression, many Indian communities and individual athletes embraced the new Canadian game of box lacrosse. Probably the most famous player was a Mohawk from Six Nations named Harry Smith. He moved to Buffalo, New York, played for semipro barnstorming teams, changed his name to Jay Silverheels, and eventually moved to California and developed an acting career. Compared to their experiences with collegiate programs in the United States, Indian athletes and reservation teams fared better under the jurisdiction of Canadian governing bodies. This is especially true for the American and Canadian lacrosse halls of fame. The Indian presence in the U.S. hall of fame is small; the Canadian hall of fame includes a much larger number of Natives, including a father and son, Ross (1969) and Gaylord (1990) Powless. The many Iroquois reservation communities of New York, Ontario, and Quebec have used both field lacrosse and box lacrosse to reinforce intertribal social ties. Many reservation communities have also continued to stage community-building games that deemphasize competition.

Recent Trends

New developments have shaped the world of lacrosse during the past three decades: the rise of the plastic stick, the emergence of professional leagues, and the creation of international championships. Before the 1970s, Iroquois wood craftsmen manufactured all lacrosse sticks. These craftsmen had small factories on reservations, including Saint Regis, Onondaga, and Six Nations. However, in 1968 a fire that destroyed the largest Native stick factory showed the broader lacrosse community how dependent the market was on the production capacity of these craftsmen. The Chisholm factory at Saint Regis recovered from the disaster, but non-Native entrepreneurs by 1970 had developed standardized, molded lacrosse stick heads made from synthetic materials. In only a few years, lacrosse players everywhere were playing with plastic sticks made by companies such as STX. During the 1980s and thereafter, companies like Brine, Warrior, deBeer, and Mohawk International Lacrosse produced a variety of factory-made synthetic equipment.

Professional lacrosse was attempted many times throughout the twentieth century, but all efforts were short-lived commercial failures until 1987, when promoters created indoor lacrosse and sold to the public their Eagle Pro Box Lacrosse League. The league embraced a single-entity ownership model—that is, it owned all of the teams. After the debut season, the league was renamed the Major Indoor Lacrosse League and very slowly expanded the number of franchises and games played. A rival group of owners forced a merger with the MILL in 1997. This reconfigured National Lacrosse League shifted to the multi-owner system found in other pro sports. The surnames of many Native athletes dotted teams' rosters: Abrams, Benedict, Bomberry, General, Henry, Jacobs, Kilgour, Lyons, Powless, Red Arrow, Schindler, and Squire. One athlete, Barry Powless, went on to coach the Rochester Knighthawks and was later named the league's vice president of lacrosse operations. Another star Native athlete, Darris Kilgour, became head coach of the Washington Power.

The third new trend mentioned above, the internationalization of the game, saw lacrosse spread beyond its traditional areas—the United States, Canada, the United Kingdom, and Australia—to continental Europe and East Asia as well. In 1967, the four traditional powers met in Toronto for the first World Games. Beginning in 1974, the countries reconvened every four years for an international championship. Similar tournaments were established for women and younger athletes. In 1987, the International Lacrosse Federation boldly admitted the Iroquois Nationals as a member. They competed in their first World Games as a distinct team against the United States, Canada, Britain, and host Australia in 1990. Inspired by the efforts of the Nationals, the National Congress of American Indians voted to endorse lacrosse as the official sport of Native North Americans.

Donald M. Fisher

See also: W. George Beers; Box Lacrosse; Iroquois Nationals Lacrosse; Lacrosse Exhibition Tours; Lacrosse Sticks.

FURTHER READING

Beers, William George. *Lacrosse, the National Game of Canada.* Montreal: Dawson Brothers, 1869.
Fisher, Donald M. *Lacrosse: A History of the Game.* Baltimore: Johns Hopkins University Press, 2002.
Green, Tina Sloan, and Agnes Bixler Kurtz. *Modern Women's Lacrosse.* Hanover, NH: ABK, 1989.
Tewaarathon [Lacrosse]. Akwesasne: North American Travelling College, 1978.
Vennum, Thomas, Jr. *American Indian Lacrosse: Little Brother of War.* Washington, DC: Smithsonian, 1994.

LACROSSE EXHIBITION TOURS

During the late nineteenth century, three exhibition lacrosse tours—of white, middle-class club members and Aboriginal Canadians—went to Britain. In a series of matches, played in England, Scotland, and Ireland, white gentlemen amateurs were pitted against Native lacrosse players from Kahnawake, near Montreal. Although the purposes of the tours, in 1867, 1876, and 1883, were different, the teams of Aboriginals were always used to attract and sustain the interests of crowds. Through costumes, ceremonies, and mock rituals, tour organizers projected a sharp distinction between what they and spectators alike perceived to be the "primitive" qualities of Natives and the civilized cultural postures of white gentlemen. From town to town and city to city, such images were invoked to secure profits, promote the new "Canadian" culture while sustaining a connection with Britain, and encourage the immigration of skilled workers and wealthy families to Canada.

The first tour, in 1867, was in part stimulated by an earlier visit to Canada by the Prince of Wales, who saw value in sporting exchanges between Britain and the colonies. Capt. W.B. Johnson of the Montreal Lacrosse Club organized the first matches as a profit-making enterprise, hoping that British spectators would come to view the Canadian Indian playing a traditional game. Sixteen players from Kahnawake joined the trip and were paid twenty-five dollars for the tour. The tour was neither widely publicized nor well attended.

Lacrosse promoter W. George Beers organized the second lacrosse trip in 1876, with the intent of demonstrating the sport as real gentlemen played it. Beers had been promoting lacrosse as Canada's national game for a decade, and he viewed the tour as a patriotic celebration of Canadian manhood. Yet he too used Aboriginal Canadians to attract interest and public attention. Once again the Montreal Lacrosse Club supplied the gentlemen members and Iroquois players from Kahnawake joined them; they set out to play sixteen matches in six weeks in Ireland, Scotland, and England. This tour was more popular than the first, attracting crowds of three to five thousand. Beers promoted images of the gentleman amateur sportsman in sharp contrast from the Native players, who were given scarlet feathers to wear in their uniform caps. Beers viewed this tour as more serious and less carnival-like than its predecessor, but between matches the Native team was induced to engage in snowshoe races on the grass and perform "war" dances to entertain the spectators.

Beers organized the trip of 1883 as well; however, this time its intent related more directly to a formalized international government policy, as opposed to a cultural display and celebration of the connection between British and Canadian physical masculinity. Members of the Montreal and Toronto Lacrosse Clubs constituted the white gentlemen-amateur team, and, once again, Native players were selected from Kahnawake. The trip was sponsored by the federal Department of Agriculture in support of Prime Minister John A. MacDonald's "National Policy"—to populate the Canadian West with immigrants from Europe. Under this policy immigration agents were stationed across Europe, and cultural programs, including world's fairs and international sport competitions and exchanges, were utilized to showcase opportunities for settlement in Canada. More than eighty immigration agents coordinated with government representatives and Beers to arrange lectures in each city. The 1883 tour was viewed explicitly as a promotional opportunity to attract immigrants. This was evident in the extensive schedule of matches organized for the two teams—they played sixty-two matches in forty-one cities in a two-month period. Once again, the symbolic message being delivered was that Canada was a civilized country where gentlemen participated in cultural activities that connected to British social ideals, and a place where Natives had been colonized and "tamed."

During the tour, the players distributed 150,000 immigration leaflets containing copies of the *Dominion Illustrated News*. The tour had little to do with the promotion of lacrosse or sport, and as in the previous tours, the Aboriginal team was positioned to serve as a marker of cultural difference and level of civilization, in contrast to the gentlemen club players.

Kevin B. Wamsley

See also: W. George Beers; Lacrosse.

FURTHER READING

Morrow, Don. "The Canadian Image Abroad: The Great Lacrosse Tours of 1876 and 1883." In *Proceedings of the 5th Canadian Symposium on the History of Sport and Physical Education.* Toronto: University of Toronto, 1982.

———. "Lacrosse as the National Game." In *A Concise History of Sport in Canada,* edited by Don Morrow et al. Toronto: Oxford University Press, 1989.

Wamsley, Kevin B. "Nineteenth Century Sports Tours and Canadian Foreign Policy," *Sporting Traditions* 13, no. 1 (1997): 73–90.

LACROSSE STICKS

Developed perhaps as early as the fifteenth century, lacrosse sticks remain a fundamental, if often overlooked, feature of the indigenous sport. Ball games now collectively referred to as "lacrosse" were played by tribal groups in the northeastern, southeastern, and Great Lake regions of North America. Regional differences in rules and play shaped stick design.

It is believed that lacrosse sticks are descendants of war clubs. Samples of several historical sticks still available for inspection have intricately carved butts similar in design to war clubs—for instance, modest representations of a hand clasping a ball alongside of which is a carving of a hand-

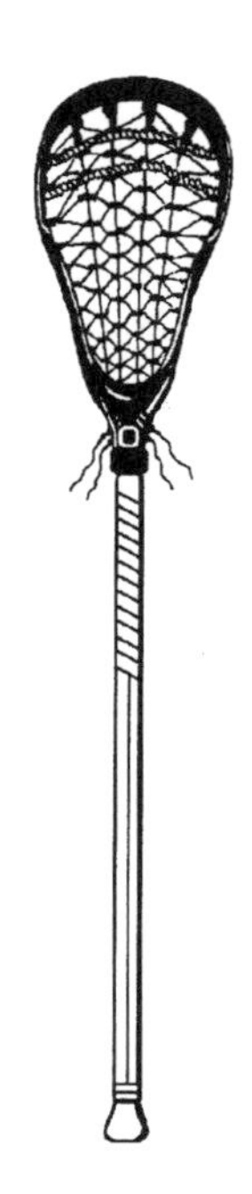

A lacrosse stick. *(Lacrosse Hall of Fame)*

shake. This is symbolic not of the friendly nature of lacrosse but rather of certain beliefs relating to battle; the handshake is representative of warriors clasping hands as protective medicine as they prepare for combat.

The first descriptions of the sticks used are by Europeans, who often made comparisons to European objects, such as racquets or bats. All known lacrosse stick designs fall into two categories: those with pockets completely enclosed by wood and those with open pockets. Generally the unenclosed stick was used by the northeast Indians; their designs varied from Great Lakes to Iroquois sticks. Native groups on both sides of the present international border utilized this stick design, one that was adopted by non-Native players in the nineteenth century when lacrosse became popular as a recreational sport in Canada and the United States. Accordingly, this became the stick design of choice and endures into the modern period.

The enclosed design, used predominantly in the Southeast, is best described as having a rounded pocket located at the end of the shaft. These sticks were used in pairs by each player; their design conforms to the southeastern lacrosse style, as played by the Cherokee, Creek, Seminole, and Yuchi, among others. The southeastern sticks were designed to cradle the ball and shoot with strength and accuracy. The northeast groups' sticks—generally longer and with much larger oval, or tear-shaped, pockets—improved players' ability to stick-handle, though they were also designed for powerful and accurate shots. Balancing and cradling the ball in this pocket design required tremendous skill. Due to the size of the pocket it was easier for opposing players to dislodge the ball. The size of the pocket of each stick design was in direct relation to the size of the ball, which could be either mostly wooden or made of slightly lighter buckskin.

Traditionally, either individual players or recognized makers produced sticks. When creating a southeastern stick, the maker selected a hickory log about four feet long, which was then split. After stripping the bark , the craftsman whittled the wood down into a sixteen-inch portion with one end rounded, which formed the cup once bent back upon itself. The shaft was then steamed and the ends turned to player specifications. Deer hide was utilized to secure the tapered end to the handle. Similarly, Great Lake sticks were made of hickory or in many cases white or black ash; the selected log was split in half, and then in half again, and then again, making available to the stick maker eight shafts approximately fifty inches in length. The shaft was then whittled and the end tapered to about one-quarter inch thick. The thin wood at the pocket was quite pliable and once steamed was curled back inside the body and secured. Once this task was completed, webbing of buckskin or leather was then woven through drilled holes to finish the pocket.

The Iroquois stick employs the same technology as the Great Lakes stick, although hickory was the wood of choice. The wood was split and whittled down to where the pocket was located, at the shaft's end. The primary difference between the Iroquois and the Great Lakes stick was in the pocket design; in the Iroquois version the wood was steamed and manipulated into shape by a bending mold. Once the stick dried into shape, it was carved to desired width and length, after which all remaining bark was removed. Four to five holes were then drilled at the top of the pocket to accept rawhide runners; this step was followed by the application of lacquer. At this point the stick was ready to receive the various strings of rawhide and catgut that made up the webbing.

As recently as a century ago, Tuscarora craftsman were still using hickory and rawhide to produce lacrosse sticks. More recently, the increasing popularity of lacrosse has changed stick design. Sticks today reflect allegiance to traditional designs, but the wooden shaft has now been replaced by plastic and titanium, and sinew and rawhide by nylon and commercial leather webbing. Accordingly the weight of the stick has also dropped, from a few pounds to twelve ounces in many cases.

Yale D. Belanger

See also: Heritage; Lacrosse.

FURTHER READING

Jones, Janine, and Pam Dewey. "Little Brother of War." *Smithsonian Magazine* (December 1997): 32–33.

Lyford, Carrie A. *Iroquois Crafts*. Stevens Point, WI: Robert C. Schneider, 1982.

Vennum, Thomas, Jr. *American Indian Lacrosse: Little Brother of War*. Washington, DC: Smithsonian, 1994.

Naomi LANG

Born December 18, 1978
Ice skater

Naomi Lang, a member of the Karuk tribe of northern California, is the only Native American woman to represent the United States in the Winter Olympics. She and her ice dance partner, Peter Tchernyshev, qualified for the 2002 Winter Olympics by winning the U.S. championship in January 2002, their record-setting fourth consecutive title. Lang was born to a Karuk father and a French/English/Irish-American mother. Naomi and her mother left northern California, the ancestral home of the Karuk tribe, when she was eight and moved to Allegan, Michigan. Naomi had

Naomi Lang and Peter Techenyshev compete in the 2003 U.S. Figure Skating Championship. *(AP/Wide World Photos)*

started ballet at age three in California. She attended an Ice Capades at age eight and decided she wanted to ice skate. With a background in ballet, ice skating was a natural adjustment for her dancing abilities. She attended her first ice skating lesson in Kalamazoo, Michigan.

At age thirteen, Lang started commuting to Detroit and began skating for the Detroit Skating Club. At age fifteen, she started living with a family in Detroit and spending her weekends in Allegan with her mother. Her first partner was John Lee, and they won their first competition at the novice level. At age eighteen, she relocated to Lake Placid and began skating with Peter Techernyshev.

After a year in New York, Lang and Tchernyshev moved back to Michigan with Lang's mother to train under Lang's old coaches, Igor Shpilband and Liz Coates. The pair placed third in the U.S. Nationals in 1998. This was followed by national championships in 1999 and 2000. Feeling that they needed something new if they were to bring their skating to a higher level, the pair moved back to the New York area. Their new coach and choreographer, Alexander Zhulin, at the Ice House in Hackensack, New Jersey, prepared them for the national championships in 2001 and 2002. In the 2002 Winter Olympics in Salt Lake City, Utah, they finished in eleventh place.

In the future, Lang hopes to provide skating instruction to Native American children and to attend college to earn a degree in veterinary medicine.

Royse Parr

FURTHER READING

Chataigneau, Gerard, and Steve Milton. *Figure Skating Now.* Willowdale, ON: Firefly Books, 2001.

Native American Times (www.okit.com/sports/2002/janfeb/naomi.html).

Reggie LEACH

Born April 23, 1950
Hockey player

Leach, born to unmarried teenage parents, grew up in Riverton, Manitoba. His father left to work in the mines before he was born, and his Cree mother soon left for Edmonton, Alberta. Leach was raised by his paternal grandparents along with twelve other children in a home beset with poverty and alcoholism. Never a good student, young Leach found his calling in hockey, becoming one of the premier goal-scoring wingers in the junior and professional game.

At the age of thirteen, he was recruited to play with adults on a semipro club. News of the talented youngster's abilities spread quickly. Leach soon joined the Flin Flon Bombers, the top junior club in Manitoba. He enjoyed great success playing on the line centered by Flin Flon's top player, Bobby Clarke.

Reggie Leach (right) celebrates after scoring his 300th career goal. *(AP/Wide World Photos)*

The Boston Bruins selected Leach in the first round of the 1970 National Hockey League (NHL) entry draft. Leach struggled early in his professional career. He took pride in his ability to shoot the puck, but his defensive play and his work habits in practice drew the ire of his coaches. Leach received scant playing time with the powerhouse Bruins team. In 1972, Boston dealt him to the lowly Oakland Seals, with whom he underachieved. After the 1973–1974 season, Leach got a chance at a new hockey life. One week after the Philadelphia Flyers won the Stanley Cup, they traded for Leach, reuniting him with Clarke.

Leach started poorly with Philadelphia. Through the first quarter of the 1974–1975 season, he scored only three goals. Receiving encouragement from Clarke and a challenge from coach Fred Shero, Leach turned around his season and his career. He scored forty-five regular-season goals and added eight playoff tallies to help Philadelphia win their second Stanley Cup. The next season, Leach was even better, scoring sixty-one regular-season goals. In sixteen playoff contests, the "Riverton Rifle" pumped home nineteen goals, a record that still stands (it was later equaled by Jari Kurri of the Edmonton Oilers). Although Philadelphia lost to the Montreal Canadiens in the Stanley Cup finals, Leach won the Conn Smythe Trophy as the most valuable player in the playoffs.

For the next several years, Leach's production was sporadic. Heavy alcohol consumption affected his performance. Nevertheless, he continued to average about thirty goals a season. Finally, Leach rebounded to score fifty goals in the 1979–1980 season, helping Philadelphia return to the Stanley Cup finals.

Leach's career began to decline shortly thereafter. He finished his NHL career with a poor Detroit Red Wings team in 1982–1983, for whom the alcoholic winger produced little. He concluded his NHL career with 383 goals. Leach's son, Jamie, also played in the NHL. After his retirement, the elder Leach conquered his alcoholism and established successful business ventures. He continues to travel to native communities in North America, warning youth of the dangers of substance abuse.

William R. Meltzer

FURTHER READING

Fischler, Stan. *The Greatest Players and Moments of the Philadelphia Flyers*. Champaign, IL: Sports, 1998.

Greenberg, Jay. *Full Spectrum: The Complete History of the Philadelphia Flyers Hockey Club*. Chicago: Triumph, 1996.

Hart, Gene, and Buzz Ringe. *Score! My Twenty-Five Years with the Broad Street Bullies*. Chicago: Bonus Books, 1990.

A.E. "Abe" LEMONS

Born November 21, 1922
Died September 2, 2002
Basketball coach

Abe Lemons won 599 games and produced several All-Americans as one of the premier major basketball coaches in the United States from 1955 to 1990. At Oklahoma City University, Pan American University, and the University of Texas, the humorous Abe Lemons became the most quoted basketball coach ever. USA Today chose Lemon's famous line "Doctors bury their mistakes, ours are still on scholarship" as the sports quote of the twentieth century. The thousands of quotable, hilarious remarks make up only a small part of the legacy of Abe Lemons. Behind the funny-man exterior was a superb basketball mind. Former Indiana University and now Texas Tech coach Bobby Knight considered Lemons as one of the five best basketball minds in the history of the game. Some have said his skill of coaching basketball players on offense was unparalleled. Often his teams were among the national leaders in scoring.

Born in poverty on a farm in southwest Oklahoma, Lemons struggled to get a college education, which was interrupted by his service as an officer in the merchant marine during World War II. After the war, he became a star basketball player and team captain for the Oklahoma City University (OCU) Chiefs. When he graduated with a degree in physical education in May 1950, he was named an assistant coach at his alma mater. He also coached the university's freshman team for four seasons. When OCU's head coach departed to coach Oklahoma University, Lemons was promoted to replace him, beginning with the 1955–1956 season.

Lemons's parents had raised him in the land of the Kiowa, Comanche, and Apache to be proud of his Indian heritage, but he never was told which tribe he could claim. He was renowned as a recruiter of overlooked, talented basketball players in the small towns of Oklahoma. Two of his prize recruits were full-blood Kiowa, Fred Yeahquo and Joseph "Bud" Sahmaunt, who led the 1958–1959 OCU team to the National Invitational Tournament. Sahmaunt, who was an honorable mention All-American in 1959, became the OCU athletic director in 1983.

Shortly after Lemons's retirement from OCU and coaching in 1990, he was inducted into the Oklahoma Sports Hall of Fame. In 1994 he was chosen as a member of the University of Texas Longhorn Hall of Fame. In 1995 the Jim Thorpe Association created an "Abe Lemons Award," to be presented annually to a member of the association's executive council who demonstrated the same qualities as Lemons. In 1998 Oklahoma City University honored Lemons by naming a new sports facility on its campus the Abe Lemons Arena.

In 1997 Lemons was inducted into the Oklahoma Coaches Association Hall of Fame. By 1998, Lemons was slowed by Parkinson's disease, a degenerative neurological disorder marked by tremors and slow movement. Accepting his induction into the Oklahoma Heritage Association's Hall of Fame in 1999, Lemons responded to questions about his tremors in a way that reflected his lighthearted outlook on life: "It's not so bad. I finally have rhythm."

Royse Parr

FURTHER READING

Burke, Bob, and Kenny Franks. *Abe Lemons: Court Magician*. Oklahoma City: Heritage, 1999.

Nelson LEVERING

Born 1926
Boxer

A member of the Omaha-Bannock tribe, Levering attended Haskell Indian Institute, where he garnered recognition as an amateur boxer and was ultimately inducted into their American Indian Athletic Hall of Fame.

After attending Haskell and fighting for a while in the amateur ranks, he turned professional and had twenty-eight fights. He fought welterweight matches during the middle part of the twentieth century, from the late 1930s to the early 1950s. At this period, welterweights fluctuated between approximately 136 and 147 pounds per fight. In 1947, Levering won the Midwest Golden Glove championship. In 1948, Levering gained prominence by becoming the Kansas State welterweight champion.

As a welterweight, Levering made several national appearances; he had the honor of appearing on four different fight cards with the esteemed heavyweight champion Joe Louis. In the second half of the 1950s Nelson Levering ended his illustrious career, finishing with an amateur record of 35–5 and a professional record of 23–5, winning seventeen fights by knockout. In 1981, Levering was honored by induction into the American Indian Athletic Hall of Fame.

Othello Harris and J.R. Wampler

FURTHER READING

"Hall of Fame/Nelson Levering" (www.ndnsports.com/index.asp), February 12, 2002.

Oxendine, Joseph. *American Indian Sports Heritage.* Champaign, IL: Human Kinetics, 1988.

George LEVI

Born 1899, Bridgeport, Indian Territory
Died Unknown
Football player

Levi (Arapaho) attended Haskell Indian Institute between 1923 and 1926, playing football and basketball and running track. On the gridiron, he played left halfback and was teamed in the backfield with his older brother, an All-American, John Levi. The *Haskell Annual* praised younger Levi's all-around talents: "He is a great passer, a fine open-field runner, a vicious tackler, and terrific lineplunger." In 1923, he earned honorable mention All-American honors.

Despite being a fine athlete in his own right, Levi lived much of his life in the shadow of his older brother. His nickname "Little Skee" demonstrated how difficult it was to escape the long shadow of his more famous brother, who played professional baseball, attempted to participate in the 1924 Olympic Games, and coached at Haskell.

William J. Bauer, Jr.

See also: John Levi.

FURTHER READING

Adams, David Wallace. "More than a Game: The Carlisle Indians Take to the Gridiron, 1893–1917." *Western Historical Quarterly* 32 (Spring 2001): 25–54.

Newcombe, Jack. *The Best of the Athletic Boys: The White Man's Impact on Jim Thorpe*. Garden City, NY: Doubleday, 1977.

Oxendine, Joseph. *American Indian Sports Heritage.* Champaign, IL: Human Kinetics, 1988.

Steckbeck, John. *Fabulous Redmen: The Carlisle Indians and Their Famous Football Teams*. Harrisburg, PA: J. Horace McFarland, 1951.

John "Skee" LEVI

Born 1898
Died 1946
Football player, baseball player

John Levi, a multisport athlete who excelled in football and baseball, is considered to be the greatest athlete ever produced by the Haskell Institute.

Levi, an Arapaho Indian, was the son of Tom Levi and his wife Cecelia Goodkiller. Before arriving at Haskell in 1921 Levi attended a succession of schools, including the Arapaho School in Oklahoma, the Chilocco Indian School, Phillips University, and the Chillicothe Business College.

The first mention of Levi in the school newspaper came in 1921, when he returned a kickoff seventy yards for a touchdown against Kansas Wesleyan. A fullback, with speed and strength as a ball carrier (at six feet, two inches and 190 pounds), Levi combined the ability to throw a forward pass up to seventy yards in the air along with outstanding punting and dropkicking skills. Soon he was being described as a triple-threat back and the second coming of the legendary Jim Thorpe.

In 1922 Levi was named team captain of the football team and scored a total of eighty-six points. During the 1923 season Levi began to achieve national renown for his football talents—helped in great part by his performances against the University of Minnesota and the Quantico Marines at New York City—and he finished the year with a total of twenty-three touchdowns. In the 1923 postseason Levi received a first-team fullback berth on the All-American squad named by *Athletic World*, along with a second-team spot from Walter Eckersall of the *Chicago Tribune*. In the fall of 1924, Levi again was the star of Haskell's backfield, scoring eighteen touchdowns and 112 points.

Levi also represented Haskell in baseball, basketball, and track. He once played in a baseball game against Drake University in which, between innings, he managed to win the shot put, discus, and high jump events in a track meet against Baker University. Levi won more athletic letters than any other individual in Haskell's history. He was considered to be a serious challenger for the U.S. Olympic track and field team in 1924, but instead he signed a baseball contract with the New York Yankees.

Beginning his professional baseball career as an outfielder in 1925, Levi was assigned to Harrisburg of the New York–Pennsylvania League, where he batted .346 with thirty doubles and ten home runs. Returning to Haskell, Levi became an assistant coach for several sports from 1926 to 1936. During this period he also served briefly as player/coach of the Hominy Indians, a semipro football team. Levi resumed his minor league baseball career in 1929 at Topeka of the Western League, where he batted .317 with thirty-six doubles and thirteen home runs, and then hit .327 with five home runs in thirteen games at St. Joseph in 1930.

Jim Thorpe once described John Levi as the greatest athlete he had ever seen. Levi has been inducted into the American Indian Athletic Hall of Fame (1972), the Oklahoma Athletic Hall of Fame (1973), and the Kansas Sports Hall of Fame (1974).

Raymond Schmidt

FURTHER READING

McDonald, Frank W. *John Levi of Haskell.* Lawrence, KS: World, 1972.

Oxendine, Joseph B. *American Indian Sports Heritage.* Champaign, IL: Human Kinetics Books, 1988.

Stucky, Tim. "Native American Legends, Levi and Weller." *Kansas Sports* (December 1996–January 1997): 12.

Walter LINGO

Born October 12, 1890
Died 1969
Football team owner

Walter Lingo established the Oorang Indians football team of LaRue, Ohio, one of the original franchises of the NFL. Lingo, born October 12, 1890, grew up in LaRue—a town with a population of 750. As a youth he was fascinated by animals and Native American cultures, and eventually he would become LaRue's wealthiest citizen as a result of a kennel business in which he bred Oorang Airedale canines, a breed that had emerged originally in England and Scotland.

Seeking adventure, as a boy he often ran off during the summers to find odd jobs; as a teenager he "skipped out west and spent the summer with the Indians." Later, he began erecting an Airedale empire, built in part on the mail-order sale of puppies. Lingo never lost his robust enthusiasm for the outdoors, and in particular for hunting wild game. He began many friendships by inviting people to join him for hunting engagements in Ohio and in other regions of the United States. Lingo's hunting partners included Warren G. Harding (from nearby Marion, Ohio), elected president of the United States in 1921, and the famous Sac-Fox athletic superstar Jim Thorpe. During his famous hunting trips, Lingo liked to chronicle miraculous tales of Airedales rescuing children or performing acts of courage on the World War I battlefield, many having been trained for rescue tasks by the Red Cross.

For Lingo, breeding and raising these hunting dogs were activities connected to the mythos of the wilderness. He believed that Native American peoples too embodied a unique relationship to nature, and thus assembling a professional team of Native American football players as a way to help sell Airedales seemed logical. Indeed, he perceived a supernatural bond between these dogs and the American Indian: "I knew that my dogs could learn something from them that they could not acquire from the best white hunters." In his mind, American Indian people were quintessentially masterful hunters and trackers; he "considered [them] to be mythic people and believed there was a supernatural bond between Indians and animals."

Lingo hired Thorpe to serve as the player-coach of the Oorang Indians franchise, which he purchased in June 1922 for a sum of $100. All of the players were American Indians, and Lingo insisted they play most of their games on the road in order to be of best use as an advertising vehicle. During halftime, players dressed in Indian regalia and performed stunts on the field with the Airedales. Moreover, Lingo asked the Native Americans to help train the dogs by running with them in the evenings along the banks of the Scioto River. Under these constraints, the team fared poorly, and by 1924, the franchise folded.

Charles Fruehling Springwood

See also: Oorang Indians.

FURTHER READING

Borowski, Jim. "Tiny LaRue Was Once an NFL Town." *Sunday Oregonian*, January 8, 1995, 1.

Whitman, Robert L. *Jim Thorpe and the Oorang Indians: The N.F.L.'s Most Colorful Franchise.* Marion, OH: Robert Whitman and the Marion County Historical Society, 1984.

LITERATURE

Native American literature abounds with references to sports. Indigenous authors often use sports to signify the civilization of Native Americans or to affirm the cultural continuity of a particular tribal tradition. They also include references to native sports to reflect on personal or communal histories, celebrate their cultures, or reclaim Native terms. From the earliest Native American texts, such as pictographic traditions like the Dakota *ozan,* Native storytellers have long been interested in portraying tribal games.

For some Native authors, sports become a way of restoring a community to wholeness. In the short story "The Only Traffic Light on the Reservation Doesn't Flash Anymore," Sherman Alexie pursues the notion of the reservation basketball star as a savior to his or her people. A pair of fallen basketball heroes sit on a front porch watching new stars emerge, fall victim to alcoholism, and ultimately fail. Neverthe-

Spokane-Couer d'Alene author and poet, Sherman Alexie has often incorporated basketball in his writings. *(AP/Wide World Photos)*

less, the community continues to invest hope in each next generation of ballplayers. Similarly, in Tomson Highway's *Dry Lips Oughta Move to Kapuskasing,* hockey emerges as a community-affirming sport that may ultimately heal the Wasaychigan Reserve, which has suffered under colonialism.

Perhaps the most famous image of sports as healing pertains to an individual protagonist, Abel in N. Scott Momaday's *House Made of Dawn.* When Abel returns home to his Tano Pueblo in 1945 after serving in World War II, his experience in mainstream society has fundamentally separated him from the continuity of Tano traditions. As part of Abel's healing and reintegration into Tano society, he joins the ceremonial runners after evil, ultimately finding his own peace through this run.

Other authors choose to heal with humor in their use of images of sports, and Gerald Vizenor spoofs Euro-American/Native-American relations in a baseball game in his screenplay *Harold of Orange.* Harold Sinseer has applied for foundation money to put coffeehouses on reservations to spawn revolutions. Gratifying the foundation board's curiosity about his nativeness, he organizes a baseball game between the Warriors of Orange and the foundation directors, in which the Native team wears shirts that say "Anglos" and the foundation members wear shirts that say "Indians." Embodying the role of trickster, Harold wears layers of alternating "Anglo" and "Indian" t-shirts, so he is able to move between teams throughout the game, muddying the foundation members' strategies.

Vizenor's depiction of baseball displays a deep-rooted ambivalence toward Euro-American sports and the culture they represent, and other authors have offered similar critiques. In *Sundown,* John Joseph Mathews's protagonist, Challenge Windsor, attends a state university on a football scholarship but becomes conflicted about his own identity as a result, his alliances between his teammates and Natives on campus being at odds. He serves in the air force and eventually returns home, plagued by the problem of integrating the different worlds he has lived in.

Traditional tribal sports are often represented in Native American writing as vehicles for communicating vital cultural information to non-Indian reading audiences. In *Indian Boyhood,* for example, Charles Eastman retells how a lacrosse game was the occasion of his receiving his adult name. In a closely matched game between two Dakota bands, the opponents wager a name, "Ohiyesa," "the Winner," to be given to the winner's choice of a young man. When Eastman's people, the Wahpetons, win, Eastman is given the name, to which he is exhorted to live up and that becomes the defining trope of his emerging masculinity.

In *My People, the Sioux*, Luther Standing Bear uses the game of lacrosse by way of portraying to his non-Indian readers Native Americans as the true Americans. In chapter three, "Games," Standing Bear includes sketches in an adapted pictographic style as a method of explaining and legitimizing Lakota life to his readership; the last of these pictographs opens a discussion of "the *real American* baseball," Lakota lacrosse, in which Standing Bear questions his readerships' understanding of American identity, through his stress upon the authenticity of the Lakota as the "real" Americans.

Other traditional sports that are represented in literature include dancing, such as Susan Power's novel *The Grass Dancer*, in which the protagonist is a female grass dancer, Pumpkin. Pumpkin's story serves as a frame to the entire novel, from her winning first place at a powwow and her tragic death soon thereafter to the resurrection of her spirit when her lover, Harlan Wind Soldier, is able to overcome her loss and rejoin the dance circle. Grass dancing comes to represent a pride in the Dakota culture and nation, a pride that will carry the people through the generations.

The aspect of sport in literature that seems to speak most to contemporary issues may be gaming. From Gerald Vizenor's representations of the Windigo in *The Heirs of Columbus* to Leslie Silko's crafting of the Evil Gambler in *Ceremony*, contemporary Native American authors consistently use the image of gambling for a range of purposes, often alluding to and refashioning older tribal traditions around gambling.

Penelope M. Kelsey

See also: Films; Media Coverage.

FURTHER READING

Alexie, Sherman. *The Lone Ranger and Tonto Fist-Fight in Heaven*. New York: Atlantic Monthly, 1993.
Donahue, Peter. "New Warriors, New Legends: Basketball in Three Native American Works of Fiction." *American Indian Culture and Research Journal* 21, no. 2 (1997): 43–60.
Eastman, Charles. *Indian Boyhood*. New York: McLure, Phillips, 1902.
Highway, Tomson. *Dry Lips Oughta Move to Kapuskasing*. Saskatoon: Fifth House, 1989.
Matthews, John Joseph. *Sundown*. Norman: University of Oklahoma Press, 1988.
Momaday, N. Scott. *House Made of Dawn*. New York: Harper, 1968.
Power, Susan. *The Grass Dancer*. New York: Harper, 1994.
Silko, Leslie. *Ceremony*. New York: Penguin, 1977.
Standing Bear, Luther. *My People, the Sioux*. Boston: Houghton-Mifflin, 1928.

J. Wilton LITTLECHILD

Born April 1, 1944
Diver, scholar, activist

Born April 1, 1944, J. Wilton Littlechild is a member of the Cree Nation from the Erminiskin Reserve, one of four Cree bands that constitute the Hobemma Reserve in Alberta, Canada. As a scholar, athlete, and advocate for indigenous rights, he is one of the most influential and dynamic Aboriginal leaders in Canada.

As a youth, Littlechild attended St. Anthony's College, a local boarding school that provided both elementary and secondary educations. After graduating, Littlechild went to the University of Alberta, where he received his bachelor of physical education degree in 1967, his master of science in physical education in 1975, and a law degree in 1976. His master's thesis focused on the athletic accomplishments of Tom Longboat, the famed Onondaga runner, and the impact of Longboat's career on Canadian sport history.

Working from his office in Hobemma, Alberta, Littlechild specializes in international indigenous rights. He was called to the bar in 1972 and appointed to Queen's Counsel in 1976. He was named honorary chief in the Cree Tribe for being the first treaty Indian from Alberta to graduate from law school. He is also the first treaty Indian to become a member of Parliament, where he served from 1988 to 1993 for the constituency of Wetaskiwin. He is a founding member of the Indigenous Initiative for Peace, an international organization of indigenous leaders, and founder of the World Resource Council of Indigenous Peoples, a United Nations agency concerned with international indigenous rights.

At university, Littlechild had demonstrated his skills as an athlete and as a sport administrator. He competed for two years on the varsity hockey team and diving teams, specializing in the one-meter and three-meter events, and was later appointed as a judge for the Canadian Intercollegiate Athlete Union (CIAU) national diving championships. In different years, he was also the manager of the varsity football and basketball team. For his outstanding athletic accomplishments, Littlechild was named national recipient for the Tom Longboat Award in 1967 and 1974. He shares the distinction of being a two-time recipient of this prestigious award with only one other athlete, Alwyn Morris, gold and bronze medalist in kayak pairs at the 1984 Olympic Games.

Throughout his life, Littlechild has helped generate support for Aboriginal sport development within Canada and abroad. He is one of the original founders of the Indian Sports Olympics (INSPOL), which helped create sport and recreation programs for First Nations peoples on and between reserves in Alberta in the 1960s and 1970s, as well as the North American Indigenous Games (NAIG), a major sport and cultural festival for the Aboriginal peoples of Canada and the United States, and is the originator of World Indigenous Nations Sport (WIN Sport), an Alberta-based Aboriginal sport organization with the objective of establishing a World Indigenous Nations Games.

In 1999 he received the Order of Canada for his efforts to help establish the NAIG and for his contributions to Aboriginal sport development in Canada. In 1999 the Indian Association of Alberta established the Willie Littlechild Achievement Awards, presented annually to Aboriginal students from Alberta who demonstrate the same qualities of commitment and excellence to athletics and academics.

Janice Forsyth

See also: North American Indigenous Games.

FURTHER READING

Champagne, Duane, ed. *The Native North American Almanac: A Reference Work on Native North Americans in the United States and Canada*. Detroit: Gale Research, 1994.

Forsyth, Janice. "J. Wilton Littlechild's North American Indigenous Games." Master's thesis, University of Western Ontario, 2000.

"J. Wilton Littlechild." *Department of Indian and Northern Affairs Canada* (www.inac.gc.ca/ch/dec/jchild_e.html).

Littlechild, Wilton. "Tom Longboat: Canada's Outstanding Indian Athlete." Master's thesis, University of Alberta, 1975.

Gene LOCKLEAR

Born July 19, 1949
Baseball player, painter

Gene Locklear (Lumbee) has significant accomplishments in two sports-related areas.

He spent ten years in professional baseball (five in the major leagues). His paintings of major athletes, teams, and games have been widely displayed and brought high sale and auction prices.

Locklear, born in 1949 in the Mount Airy community (near Pembroke, North Carolina), is the son of Lonnie J. and Catherine Locklear. He was the first Lumbee to play major league baseball. His professional career began in 1969 with minor league play in Tampa, Florida. His major league career (1973–1977) as an outfielder and left-handed batter was played with the Cincinnati Reds, San Diego Padres, New York Yankees, and lastly the Nippon Ham Fighters (Japan). His best year in the major leagues was 1975, when he played a hundred games with the San Diego Padres and had a batting average of .321. In his 272 major league games, he had a .274 batting average, including 150 pinch-hit at-bats. In the minors, he won batting championships with the Eastern League's Three Rivers (Quebec) team in 1971 and the American Association's Indianapolis team in 1972.

In high school, Locklear took correspondence courses from the Minneapolis Art School, earning a degree in commercial art. While playing baseball he drew and painted in the off-season. When he retired from baseball, he increased his focus on painting, specializing in sports and Native American themes. He has painted major athletes in all sports, including Walt Frazier, Gordie Howe, Catfish Hunter, Bart Starr, Pete Rose, Kareem Abdul-Jabbar, and Jack Nicklaus. His works hang (or have hung) in the Gerald Ford White House, Smithsonian Institution, restaurants owned by Pete Rose and Johnny Bench, and the Thackeray Gallery in Hillcrest, California.

Locklear has painted murals, including one for the Baltimore Orioles's stadium, Oriole Park at Camden Yards. His 1993 Super Bowl mural, painted in front of the crowd, was auctioned for twenty-five thousand dollars. He was commissioned in 1993 to paint a limited-edition set (called the Locklear Collection) of nine famous-player baseball cards and a montage card for the Ted Williams series. In 1995 he signed a contract with Turner Sports to provide live paintings of scenes from the NBA playoff games. Each day Locklear painted a highlight scene from the game that would be shown during transitions to and from commercials in the next day's game coverage.

Glenn Ellen Starr Stilling

FURTHER READING

Hass, Bill. "Gene Locklear: Artist . . . and Baseball Player." *Carolina Indian Voice,* April 6, 1978, 1, 5.

Locklear, James. "Former Baseball Player Swings Soul into Painting." *Fayetteville (North Carolina) Observer,* July 6, 2001 (www.fayettevilleobserver.com).

Kyle Matthew LOHSE

Born October 4, 1978, Chico, California
Baseball player

Lohse, a professional baseball player, pitched for the Minnesota Twins beginning in 2001. He is a member of the Nomlaki Wintun tribe of California.

Lohse was recognized as an exceptional athlete at an early age in Hamilton, California. He started as a freshman on the Hamilton Braves high school baseball team. Four years later, he was the quarterback of the football team and a star basketball player and was voted the school's athlete of the year. Up to fifteen major league scouts attended his high school baseball games, tracking his ninety-plus miles per hour fast ball with their radar guns. Lohse's father, Larry Lohse, was a pitcher in minor league baseball for the Detroit Tigers. His uncle, John Lohse, played minor league baseball for the Philadelphia Phillies. His younger brother Erik became a promising minor league pitcher for the Minnesota Twins in 2001.

While he was attending California's Butte Community College, Kyle Lohse was drafted by the Chicago Cubs in the twenty-ninth round of the 1997 free agent draft. The Minnesota Twins acquired Lohse in a surprising trade with the Cubs in May of 1999. Lohse was in Class A ball in the Florida State League at the time and had no idea he was being mentioned in trade talks involving an established major leaguer.

Lohse spent the entire 2000 season with New Britain, Connecticut, where he won only three games and lost eighteen. He persevered through a fourteen-game losing streak. That fall, he rebounded with a three-win and no-loss record for Grand Canyon in the Arizona Fall League.

Lohse began the 2001 season with the Class AAA Edmonton Trappers of the Pacific Coast League. In mid-June, he was named as the league's pitcher of the week. The Minnesota Twins, then in the heat of a pennant race, added Lohse to their roster. After getting a no decision in his major league debut on June 22, 2001, Lohse won his next three games, capped off by a 13–5 win in Milwaukee on July 12. Like many rookies, he then saw his season hit the wall; he lost his next five decisions as the Twins fell out of pennant contention. He finished the season with a four-win, seven-loss record.

During the 2002 season Lohse was a dependable pitcher throughout a storybook season for the Minnesota Twins, winning thirteen games while losing eight. The Twins won the Central Division of the American League; they were finally eliminated in the playoffs by the eventual World Series champion Anaheim Angels. In the playoffs, he pitched in relief in three games and allowed no runs.

Royse Parr

FURTHER READING

Goddard, Joe. "Minnesota Twins: Selig May Want to Contract Them, But Twins Won't Disappear from Central Race." *(Charlotte, NC) Street & Smith's Baseball,* February 2002.

Krekel, Steve. "From Butte to the Bigs." *Chico (CA) News & Review,* October 4, 2001.

Thomas LONGBOAT

Born 1888, Ohsweken Reservation, Ontario, Canada
Died 1949
Runner

Tom Longboat, an Onondaga Indian, was born in 1888 (the actual date of birth is unclear) on the Ohsweken reservation near Brantford, Ontario, Canada. At his death, after a long illness, aged sixty-one, the tall Canadian was revered as a national folk hero and ethnic icon. A *New York Times* obituary (January 11, 1949) described him as, in his prime, the "greatest runner of them all."

Throughout his career, Longboat ran whenever the opportunity presented itself. He won his first race over a distance of five miles at the Caledon Fair. In 1906, he ran his first genuinely long-distance race on Thanksgiving Day. He completed the twenty-mile course in one hour, forty-nine minutes and twenty-five seconds.

In 1907 Longboat ran his most sensational race, at the Boston Marathon. He won the race with a time of 2:24.20. While the Boston Marathon distance at the time was only twenty-four miles, 1,232 yards and not the official Olympic distance of twenty-six miles, 385 yards that it is today, Longboat's "marathon" record was not bested at the Olympics until Emil Zatopek

Tom Longboat, celebrated as one of Canada's greatest distance runners, won the Boston Marathon in 1907. *(Boston Public Library)*

ran two hours, twenty-four minutes in the 1952 Helsinki Olympics.

At the 1907 Boston Marathon Longboat raced for the Young Men's Christian Association track club of Toronto. Shortly after this he was suspended by the YMCA for the consumption of alcohol. He joined the Irish-Canadian track club and set his sights on the 1908 London Olympics. In his buildup to the Olympics Longboat had a succession of race victories at distances from twelve to fifteen miles. His 1907 Boston performance made him a clear favorite for the 1908 Olympics, and early on the Canadian entrant took a long lead. Sadly, Longboat suffered sunstroke and languished, dropping out of the race due to heat exhaustion.

Following his Olympic experience, Longboat rebounded after joining a professional troupe of athletes. In the coming years, he competed against the best runners in the world, including many of those who had finished ahead of him in the 1908 London race. In arenas and stadiums as far afield as the Madison Square Garden in New York and the Powerhall Stadium in Edinburgh, Scotland, Longboat convincingly defeated the world's greatest distance runners including the Italian Doranado Pietri and the British champion runner Alfred A. Shrubb, who was the first distance runner to train systematically using recognizably modern conditioning methods.

At the end of his life, the spare Onondagan runner worked as a laborer with the Toronto street-cleaning department. Despite having won more than seventeen thousand dollars as a professional runner, a princely sum in the opening years of the nineteenth century, at his death the six-foot Canadian was destitute and was saved from homelessness only by the ownership of a house given to him by a grateful homeland (Canada) for winning the 1907 Boston Marathon.

He was admitted to Canada's Sports Hall of Fame in Toronto as a national who had "achieved excellence in sports." Longboat is also a member of the Canadian Amateur Athletic Hall of Fame (1960).

Scott A.G.M. Crawford

FURTHER READING

Kidd, Bruce. *Tom Longboat*. Ottawa, Ontario: Fitzhenry and Whiteside, 1980.

"Obituary." *New York Times*, January 11, 1949, 31.

Bud LONGBRAKE

Born 1963
Rodeo rider

Bud Longbrake of Dupree, South Dakota, and the Cheyenne River Sioux Tribe is known as one of the best Native American rodeo riders. In keeping with his family's rodeo tradition, Longbrake began competing at the age of nine, in 4-H rodeos. He then began competing at higher levels in both bull riding and team roping; in 1979 he decided to concentrate on riding saddle broncs. His determination to excel at saddle bronc riding was spurred on by his admiration of local cowboy Jeff Knight, who died in 1978. To honor his memory, Knight's family commissioned a buckle to be awarded to the winning saddle bronc rider at the Dupree Regional High School Rodeo. Longbrake trained constantly with the single goal of winning that buckle, and he won it at age sixteen—the start of a lifelong career.

His career moved ahead when in 1985, at the age of twenty-two, Longbrake purchased his rookie Professional Rodeo Cowboys Association (PRCA) card. With the arrogance of a rookie, he thought he would win all thirty rodeos he entered; instead he went home broke. Instead of giving up, Longbrake became even more determined and continued to focus on training with the goal of going to the National Finals. In 1990, Longbrake reached them. Although he entered in thirteenth place, he was the only rider to cover all ten horses, which means that he won not only the average but the national championship title as well. Between that win and December 2001, Longbrake reached the finals four times. The end of the 2001 season was plagued with injuries, including separations of both his right and left shoulders in different rodeos. Between surgery to repair his left shoulder and the following rehabilitation, Longbrake missed much of the 2002 season.

Once his shoulder healed, Longbrake qualified for the 2002 Olympic Command Performance Rodeo, an all-star event held in conjunction with the 2002 Winter Olympics in Salt Lake City, Utah. Longbrake and fellow Cheyenne River Sioux and team captain Tom Reeves were part of a five-man team that bested rival Team Canada in all three rounds. Wearing number twenty-three, he placed tenth in rounds one and two and seventh in round three, and also won the saddle bronc semifinals.

Beth Pamela Jacobson

FURTHER READING

Clausen, DB. "Bud Longbrake." *Eagle Butte News* (www.sioux.org/buds_story.htm), 2002.

Dyck, Ian. "Does Rodeos Have Roots in Ancient Indian Traditions?" *Plains Anthropologist: Journal of the Plains Anthropological Society* 41, no. 157 (1996): 205–19.

Danny "Little Red" LOPEZ

Born July 6, 1952
Boxer

Featherweight world boxing champion Danny Lopez was one of seven children born to a Mission Indian father and a Ute and Irish mother. Lopez grew up in a two-bedroom house with only a wood-burning stove for heat on a Ute Reservation in Fort Duquesne, Utah. When Lopez was young his father abandoned the family. For a while the family survived on welfare. Even when very young Lopez helped supplement the family's diet by hunting small game with a bow. Eventually, Lopez's mother was forced to put the children into foster care. Danny, sister Carol, and brother Larry went to live with a family in Jensen, Utah.

During his time with the foster family Lopez was beaten, sometimes severely. When he was thirteen he decided to fight back and spent a month in jail, even though the charge, assault and battery, was dropped. Once out of jail Lopez went to live with an aunt and uncle, who tried to force him to become a Jehovah's Witness. Having to defend himself his whole life made Lopez a hard hitter who could also take a punch. By the time he was sixteen, he was fighting amateur bouts.

At 130 pounds the red-haired, freckled Lopez earned a reputation as a slow starter but also a very hard and fast hitter who used both fists equally well; he knocked people out with a left hook, a left jab, or his right. In 1971 he was fighting professionally and winning. In 1976 Lopez traveled to Ghana, Africa, where he trained in

Danny Lopez, here fighting Sean O'Grady, defended his featherweight title eight times between 1977–1980. *(AP/Wide World Photos)*

the tropical heat, suffered intestinal difficulties from the food, and stayed in a hotel without hot water in order to prepare to fight featherweight champion David Kotey for the title. Despite the hardship, Lopez went all fifteen rounds and defeated Kotey. Between 1977 and 1980 Lopez defended his title eight times, including ring classics against Salvador Sanchez and Mike Ayala. Overall, he knocked out thirty-nine of his forty-eight opponents for an 85 percent knockout rate, losing only six fights. He lost the title in 1980 to Salvador Sanchez and retired six months later. In 1992 Lopez staged a comeback against Jorge Rodriguez but did not last a full round.

When he retired Lopez had earned a career $1.5 million, but he avoided the trappings of fame and wealth. He and his wife Bonnie bought a small house in San Gabriel Valley and now have three sons.

In 1987 Lopez was inducted into the World Boxing Hall of Fame. Two years later, Lopez won the Rochester Boxing Hall of Fame Special Courage Award.

Lisa A. Ennis

FURTHER READING

Jones, Robert F. "Two Hitters Who Can't Miss." *Sports Illustrated,* March 19, 1979, 16–19.

Newman, Bruce. "You Can't Keep a Good Man Down." *Sports Illustrated,* February 12, 1979, 29–31.

Wertheim, Joe. "Catching Up With. . . ." *Sports Illustrated,* September 22, 1997, 11.

Dwight LOWRY

Born October 23, 1957, Pembroke, North Carolina
Died July 10, 1997
Baseball player and coach

Dwight Lowry (Lumbee) played as a catcher and, later coached baseball for the Detroit Tigers and some of their minor league teams.

Lowry was born in Pembroke, North Carolina, on October 23, 1957. His father is Marvin Lowry. He played three sports at Pembroke High School. At the University of North Carolina at Chapel Hill, he played baseball from 1977 to 1980 and was named to the All-American team.

Lowry was drafted by the Detroit Tigers in 1980. He played eleven seasons of professional baseball (four in the major leagues). His 162-game-season batting average in the major leagues was .273; he also had career totals of five home runs and twenty-six runs batted in. He was a backup catcher for the Detroit Tigers in 1984 when they won the World Series. His best major league season was 1986, when he played fifty-six games for the Tigers. That year he was brought up as a replacement catcher for Lance Parrish, who had back problems. Lowry batted .307 and had eighteen RBIs. Prior to 1986, he had been offered a coach-

Baseball player, Dwight Lowry. *(AP/Wide World Photos)*

ing position with a Triple-A team, then a player-coach position with a Double-A team. He rejected both offers, hoping to play in the majors again.

It was expected that Lowry would be the Tigers' starting catcher in the 1987 season, but Matt Nokes and Mike Heath ended up as starters instead. Lowry's last playing year was 1990, with the Montreal Expos' Triple-A farm team in Indianapolis. Then he coached the Fayetteville Generals, a Class A South Atlantic League team (now known as the Cape Fear Crocs) from 1991 to 1993. In 1994 he was promoted to manager of the Generals. In 1995, the team had the best overall record in their league.

The Detroit Tigers moved Lowry to New York to manage their minor league team, the Jamestown Jammers. On July 10, 1997, Lowry died of a heart attack at age thirty-nine. The Tigers renamed their Player Development Man of the Year Award (which Lowry won in 1996) for him.

Glenn Ellen Starr Stilling

FURTHER READING

Friedlander, Brett. "Ex-Generals Manager Dies." *Fayetteville Observer,* July 11, 1997.

Lapointe, Joe. "Tiger Rookie Lowry: He's Still a Quiet Guy." *Detroit Free Press,* May 21, 1984, 1D.

Oren LYONS, JR.

Born 1930
Lacrosse player

Oren Lyons, Jr., a Turtle Clan Faithkeeper of the Onondaga Nation, was born in 1930. He grew up playing many sports, including lacrosse, which the Onondaga call *guh jee gwah ai* (bumping hips).

Lacrosse is more than just a game to the Haudenosaunee (People of the Longhouse) but a spiritual tradition that is woven into the fabric of their culture. "Lacrosse is the life blood of the Six Nations," says Lyons. The Haudenosaunee believe lacrosse is a gift from the Creator and is played for the Creator's enjoyment and to retain harmony. Lacrosse not only strengthens the physical body but strengthens the mind and spirit.

Oren Lyons, Jr., walks with activist Angela Davis outside the siege of Wounded Knee, South Dakota, March 1973. *(AP/Wide World Photos)*

Lyons learned his lacrosse goalie defense skills from his father Oren Lyons, Sr. In 2002, Lyons, Sr., Lyle Pierce, Sr., Stanley Pierce, Sr., and Irving Powless, Sr., were inducted posthumously to the Upstate New York chapter of the U.S. Lacrosse Hall of Fame. These men, known as the Fabulous Four, had passed on their knowledge and fire for lacrosse to the younger Lyons generation.

Lyons was a Syracuse University All-American Goalie from 1956 to 1958. After graduating from the university, he played club lacrosse. From 1959 to 1965 he played for the New York Lacrosse Club. From 1966 to 1970 he played for the New Jersey Lacrosse Club, and from 1970 to 1972 he played for the Onondaga Athletic Club. In 1988, he was inducted to the Syracuse Sports Hall of Fame. In 1991, Lyons was inducted into the Upstate New York chapter of the U.S. Lacrosse Hall of Fame.

Lyons is founder and executive of the Iroquois National Lacrosse Team. This team is unique as the only American Indian team sanctioned to compete, in any sport, internationally. The Iroquois National Lacrosse Team is composed of citizens of the Haudenosaunee Confederacy. The Iroquois Nationals are a world-class team, which they proved in 1998, when they beat England 10–9 in the World Games in Manchester, England.

In many ways, the Iroquois Nationals and Lyons are ambassadors of the Haudenosaunee, for they breathe the spiritual aspects of the game back into the competition. When they are not practicing or traveling worldwide, they are traveling to different native communities conducting clinics to keep the tradition alive and strong.

Before returning to the Onondaga Nation, Oren Lyons was a commercial artist in New York City. He illustrated a series of children's books on Native Americans. Lyons also has authored numerous books and articles, including *Dog Story*, a children's book, and *Exiled in the Land of the Free*.

Lyons is currently the political co-chair of the Haudenosaunee Environmental Task Force (HETF). A task force created by the Haudenosaunee Confederacy and composed of delegates chosen by each of the nations to work toward identifying environmental problems in their communities and working to find solutions to them. Lyons is also a professor of American studies at the State University of New York at Buffalo. Oren Lyons, Jr., is a spiritual and political leader. He has attended numerous United Nations meetings in the United States and in Geneva. He along with other Haudenosaunee leaders have traveled internationally advocating peace, environmental justice, and indigenous rights.

Barbara A. Gray

See also: Iroquois Nationals Lacrosse.

FURTHER READING

Lyons, Oren, and John Mohawk. *Exiled in the Land of the Free*. Santa Fe, NM: Clear Light, 1992.

GENERAL INDEX

SPORTS INDEX